# Notes in the Category of C

# Notes in the Category of C

## Reflections on Laboratory Animal Care and Use

Steven M. Niemi

ACADEMIC PRESS

An imprint of Elsevier

Academic Press is an imprint of Elsevier
125 London Wall, London EC2Y 5AS, United Kingdom
525 B Street, Suite 1800, San Diego, CA 92101-4495, United States
50 Hampshire Street, 5th Floor, Cambridge, MA 02139, United States
The Boulevard, Langford Lane, Kidlington, Oxford OX5 1GB, United Kingdom

**Notices**
Knowledge and best practice in this field are constantly changing. As new research and experience broaden our understanding, changes in research methods, professional practices, or medical treatment may become necessary.

Practitioners and researchers must always rely on their own experience and knowledge in evaluating and using any information, methods, compounds, or experiments described herein. In using such information or methods they should be mindful of their own safety and the safety of others, including parties for whom they have a professional responsibility.

To the fullest extent of the law, neither the Publisher nor the authors, contributors, or editors, assume any liability for any injury and/or damage to persons or property as a matter of products liability, negligence or otherwise, or from any use or operation of any methods, products, instructions, or ideas contained in the material herein.

**Library of Congress Cataloging-in-Publication Data**
A catalog record for this book is available from the Library of Congress

**British Library Cataloguing-in-Publication Data**
A catalogue record for this book is available from the British Library

ISBN: 978-0-12-805070-5

For information on all Academic Press publications visit our
website at https://www.elsevier.com/books-and-journals

Working together
to grow libraries in
developing countries

www.elsevier.com • www.bookaid.org

*Publisher:* Sara Tenney
*Acquisition Editor:* Sara Tenney
*Editorial Project Manager:* Fenton Coulthurst
*Production Project Manager:* Mohanapriyan Rajendran
*Designer:* Victoria Pearson

Typeset by TNQ Books and Journals

# Dedication

To technicians everywhere who provide daily care to laboratory animals. Your contributions to animal welfare and science are too often overlooked and underappreciated.

# Contents

# Biographical Sketch

Steven M. Niemi is Director of the Animal Resources, and Lecturer in the Department of Organismic and Evolutionary Biology, Harvard University Faculty of Arts Sciences, Cambridge, MA. With over 35 years' experience in biomedical research and commercial biotechnology, he has held senior management positions in contract drug and device development, biotech start-ups in human gene therapy and food animal genomics, and laboratory animal care and assurance. Dr. Niemi is a Diplomate and past president of the American College of Laboratory Animal Medicine and has served on numerous boards and national task forces addressing medical product development and laboratory animal welfare. He earned an AB in biology from Harvard College, a DVM from Washington State University, and then received a US Public Health Service National Research Service Award while a Postdoctoral Fellow in the Division of Comparative Medicine at the Massachusetts Institute of Technology. He later completed the Program for Management Development at the Harvard Business School.

# Foreword

This is an unconventional book about a subject conventionally laden with conflicting values. Reason versus emotion. Hope versus despair. Passion versus indifference. Knowledge versus ignorance. Certainty versus doubt. Dialogue versus distrust. Civility versus criminality. Good versus evil. Even angelic versus satanic. Each of these conflicts may be accompanied by proclaimed or actual life-or-death outcomes. Each of these conflicts also comes with inconsistencies by various parties (advocates vs. combatants) that make it easy to become confused or disgusted by the various claims and counterclaims.

This book's unconventional nature is perhaps easiest to describe by what the book is not. It is not, per se, a defense of animal research.[1] The reader is referred to Adrian Morrison's eloquent and personal contemplation, written after he was savaged by animal rights extremists, about the continuing need for animal research for the foreseeable future.[2] Nor is the purpose of this book to describe the ethical and regulatory frameworks within which laboratory animals are used in the United States today. That aspect of animal research is well covered in Larry Carbone's overview.[3] Conversely, this book is not an exposé or apologia when animals are used in research. It's not an instruction manual or textbook for animal care personnel, veterinarians, scientists, or regulators. It's not a review of recent scientific literature about laboratory animal biology or medicine. It's not a primer for lay readers. It's not a memoir. It's not fiction.

Instead, this book is a collection of essays I wrote about a field that is rapidly evolving in some aspects and glacially slow to change in others. It's sharing my thoughts, impressions, and ideas for others to digest who are already conversant about the subject; if these essays also provide illumination to those unacquainted with the details and nuances of animal research, so much the better.

---

1. The phrase "animal research" will serve as shorthand throughout this book for the use of live, captive, sentient vertebrates and higher invertebrates in research, testing, and education.
2. Morrison, A., 2009. An Odyssey With Animals: A Veterinarian's Reflections on the Animal Rights & Welfare Debate, Oxford University Press, New York.
3. Carbone, L., 2004. What Animals Want: Expertise and Advocacy in Laboratory Animal Welfare Policy, Oxford University Press, New York.

What's unconventional about this book is its content rather than its format.[4] I'm not aware of any other publications, with an emphasis on "public," that delve into what it's like to be embedded in the practice and enablement of animal research. Questions such as why am I doing this, what's rewarding about it, what's wrong with it, has my career been worthwhile, etc. are customarily not raised with outsiders. Instead we usually echo the party line that the responsible use of laboratory animals continues to be a vital means of advancing human and animal health (which is still true, by the way) and say no more. Those of us whose job it is to provide care and veterinary support to laboratory animals usually don't reveal our feelings, concerns, misgivings, and suggestions for improvement to each other, much less in a medium for public consumption. Even the act of identifying oneself to lay persons as a participant in animal research is usually avoided, including gatherings with friends and neighbors. This allows us to avoid arguments that usually lead to nowhere or being targeted by zealots, especially in this age of hypersocial media and anonymized threats.

So why stick one's head out of the foxhole and risk enemy fire? Simply because the conflicting values listed above in the first paragraph are so emotionally challenging and intellectually fascinating that they deserve more airtime. Not for the purpose of trying to establish any superiority of one viewpoint over another, but rather to enrich a neglected conversation that can advance the field in so many ways. To that end, this book is shaded gray rather than black and white, and devoted more to possibilities than absolutes. In laboratory animal regulatory compliance speak, it's about "could" rather than "should" or "must." Some of my statements may be deemed eccentric, some may be discomforting, some may be old news to those more knowledgeable than me; hopefully, no statements are inaccurate while all are sincere and intended to provoke thought, not histrionics. On that last point, all opinions expressed are mine alone and may not represent those held by current or past employers, professional organizations to which I belong, colleagues, family, or anyone else. I have no commercial or financial conflicts to declare; vendors are named only for example purposes, and their mention does not imply an endorsement of any products or services.

There is much jargon in the book that is familiar to those fluent in the subject matter. To make for less cumbersome text and easier reading, such jargon

---

4. Essays are compartmentalized in four distinct but related areas of animal research: Care (husbandry), Management, (veterinary) Medicine, and Ethics. I've tried to minimize overlap and redundancy but the reader should appreciate that each of these areas affect and are affected by the others. In addition, regulatory compliance, whether by internal (institutional) or external (federal, state, and municipal) means, is always lurking in the background as sometimes a determinant and sometimes an influence on the four areas under which the book is organized. The term "program" also deserves definition; throughout this book it means an institutional program of laboratory animal care, usually comprising husbandry and veterinary staff, departmental administrators, and senior management (directors). A given program may also perform its own discovery research, usually in the realm of laboratory animal biology and medicine, but its primary role is to provide animal care support to researchers at their institution.

will not be defined or explained each time it's used in each chapter; a list of abbreviations is provided to assist those who may find it helpful. Similarly, two references are so universally relevant throughout the book that they are cited in full only here rather than repeatedly cited in various chapters, i.e., the ILAR Guide to the Care and Use of Laboratory Animals (henceforth referred to as "the Guide") and Russell and Burch's seminal work on the "3Rs" (Replacement, Reduction, Refinement).[5]

This leads us to the book's title. It's an intentional word play on a commonly used list of animal pain and distress categories taken from the Annual Report form that each registered research facility is required to submit to the United States Department of Agriculture (USDA)[6] under the federal Animal Welfare Act (AWA).[7] That report (Form 7023) must list each regulated species used over the past year, how many animals of each species were used (to be indicated in Column A), and to which pain and distress category (Columns B–E) each animal should be assigned and then totaled. Definitions of each category (Column) are quoted from Form 7023 as follows:

- "Column B—number of animals being bred, conditioned, or held for use in teaching, testing, experiments, research, or surgery but not yet used for such purposes;
- Column C—number of animals upon which teaching, research, experiments, or tests were conducted involving no pain, distress, or use of pain-relieving drugs;
- Column D—number of animals upon which experiments, teaching, research, surgery, or tests were conducted involving accompanying pain or distress to the animals and for which appropriate anesthetic, analgesic, or tranquilizing drugs were used;
- Column E—number of animals upon which teaching, experiments, research, surgery, or tests were conducted involving accompanying pain or distress to the animals and for which the use of appropriate anesthetic, analgesic, or tranquilizing drugs would have adversely affected the procedures, results, or interpretation of the teaching, research, experiments, surgery, or tests."

This categorization scheme and minor variations have also been applied by many institutions and their animal oversight committees to species beyond

5. Institute for Laboratory Animal Research, National Research Council, National Academy of Science, 2011. Guide for the Care and Use of Laboratory Animals, eighth ed. National Academies Press, Washington. Available at: http://www.nap.edu/catalog.php?record_id=12910; Russell, W.M.S., Burch, R.L., 1959. The Principles of Humane Experimental Technique. Methuen, London. Available at: http://altweb.jhsph.edu/pubs/books/humane_exp/het-toc.
6. Research Facility Annual Reports. United States Department of Agriculture Animal and Plant Health Inspection Service. Available at: https://www.aphis.usda.gov/aphis/ourfocus/animalwelfare/SA_Obtain_Research_Facility_Annual_Report.
7. Animal and Plant Health Inspection Service, United States Department of Agriculture, 2017. Animal Welfare Act and Animal Welfare Regulations (Blue Book). Available at: https://www.aphis.usda.gov/animal_welfare/downloads/AC_BlueBook_AWA_FINAL_2017_508comp.pdf.

those covered under the AWA. C is the least painful or distressful Category for animals actually used in research. It is in the spirit of trying to invoke the least harm that this book is written.

I don't claim to be nor should I be considered an expert in any of the scientific, legal, or ethical realms covered in this book. But I've been immersed in all these elements simply by default, starting with a summer job taking care of laboratory rats and mice during high school, followed by majoring in biology in college, then being a laboratory animal technician, veterinary student, postdoctoral fellow (resident) in comparative medicine, board-certified specialist in laboratory animal medicine, biotech business executive, and finally director of animal care or animal welfare assurance at major academic research institutions. It's from all these experiences and from enriching conversations with so many others that inspired me to write.

To that end, I happily acknowledge and thank the following individuals who provided helpful information and insight for this book: Liz Bankert, Taylor Bennett, Jennifer Camacho, Joy Cavagnero, Randy Gollub, Claudia Harper, Donna Jarrell, Jason Jorgenson, Alla Katsnelson, Hilton Klein, Julie Lane, Chris Lawrence, Ben Lewis, David Morton, Chris Newcomer, Alyssa Terestre Pappa, Kate Pritchett-Corning, Fernando Quezada, Joan Rachlin, Andrew Rowan, Bo Rueda, Josh Sanes, Sai Tummala, and Mike Toon.

Beyond the book, I've been blessed with terrific mentors and advocates throughout my life who helped guide my career or generously provided their time and advice for my betterment. Thanks to John Gorham for encouraging an interest in veterinary medicine from the beginning; Leo Bustad and Frank Loew on how compassionate care of animals can be compatible with animal research and how to balance these two endeavors successfully in large institutions (in their case, veterinary schools); Jim Fox on how a deep dive into laboratory animal medicine can be a fulfilling profession; Bob Taber on how to succeed in business while enjoying the ride; Phil Senger and Jim Evermann on the thrill of scientific discovery and conveying that excitement to others. Finally, to my wife, Jan, to whom I owe everything.

**Steven M. Niemi**
Cambridge, Massachusetts

# Abbreviations

| | |
|---|---|
| **AAALAC** | Association for the Assessment and Accreditation of Laboratory Animal Care, International (recently and officially abbreviated to just "AAALAC") |
| **AALAS** | American Association for Laboratory Animal Science |
| **ACLAM** | American College of Laboratory Animal Medicine |
| **APHIS** | Animal and Plant Health Inspection Service |
| **AVMA** | American Veterinary Medical Association |
| **AWA** | US federal Animal Welfare Act |
| **CRISPR** | Clustered Regularly Interspaced Short Palindromic Repeats |
| **CRO** | Contract Research Organization |
| **EPA** | US Environmental Protection Agency |
| **FDA** | US Food and Drug Administration |
| **FTE** | Full-time equivalent (employee) |
| **HVAC** | Heating, ventilation, and air conditioning system |
| **IACUC** | Institutional Animal Care and Use Committee |
| **ILAR** | Institute for Laboratory Animal Research |
| **IVC** | Individually ventilated cage |
| **MGH** | Massachusetts General Hospital |
| **NHP** | Nonhuman primate |
| **NIH** | National Institutes of Health |
| **OLAW** | Office of Laboratory Animal Welfare, US National Institutes of Health |
| **PPE** | Personal protective equipment |
| **PRIM&R** | Public Responsibility in Medicine and Research |
| **SCAW** | Scientists Center for Animal Welfare |
| **SPF** | Specific pathogen-free |
| **The Guide** | The Guide for the Care and Use of Laboratory Animals, eighth edition (2011). Institute for Laboratory Animal Research, National Research Council, National Academy of Sciences |
| **USDA** | United States Department of Agriculture |

# Part I

# Laboratory Animal Care

# Chapter 1

# Must Ad Libitum Mean "Always Available"?

A trait shared by many laboratory animals and humans is the fact that if more food is readily available, it will likely be consumed, regardless of the body's need for those additional nutrients. And consuming more food than is needed can lead to obesity, type 2 diabetes, and other bad outcomes. Conversely, experimental evidence has shown for decades that reducing dietary calories to 25%–40% below what's considered normal can significantly extend an animal's lifespan and forestall many adverse conditions associated with older age, such as hypertension, cancer, cataracts, and immune-mediated diseases. These benefits of caloric restriction have been demonstrated in nematodes, fruit flies, mice, rats, and dogs, to name a few.[1,2] Rhesus monkeys were added to this list very recently after earlier studies yielded conflicting findings[3] while human trials are still ongoing.[4]

During the 1990s, regulatory toxicology found itself in a bit of a pickle due, in part, to ad libitum (over)feeding of rodents intended as test subjects. Commercial breeders had been selecting the larger siblings in generations of litters of rats or mice under the assumption that the bigger, the more robust and in more demand, and that smaller animals would be considered runts or faulty and not acceptable for regulatory data scrutiny. This preference naturally resulted in ever faster growing animals that occupied more cage space sooner, which wouldn't have been a problem for studies of 6 months or less. But chronic toxicity and carcinogenicity assays involving frequent administration of a chemical (commonly known as a test article) to rodents for 1–2 years led to earlier appearances of "background" chronic diseases and tumors, as well as earlier euthanasia, that likely had nothing to do with the effects of the chemical. This in turn threatened the scientific rigor of comparing lesions between groups of various dose levels and the no-dose group to determine if the test chemical was truly toxic or the cause of cancer in these animals. If the no-dose group had too many lesions due to accelerated aging from excessive weight or body fat or had insufficient numbers of animals remaining at the end of the study, then how could health safety regulators determine whether lesions in the dosed groups were from the chemical or not? The solution instituted by the National Toxicology Program was two-fold: to instruct commercial rodent breeders to supply federally funded toxicity and carcinogenicity

Notes in the Category of C. https://doi.org/10.1016/B978-0-12-805070-5.00001-1

**3**

studies with animals of average size rather than of large size and to change the prescribed base diet to a lower protein and higher fiber version.[5,6] It may not have been possible to reduce the food intake of animals as an alternative approach even though the health advantages of caloric restriction in animals had been published years prior. That's because many of these long-term studies involved dosed feed where the chemical being evaluated was ground up and mixed with powdered food to ensure animals were ingesting enough chemical on a frequent (daily) basis. Limiting food intake could risk limiting test article intake and perhaps not crossing an important toxicological threshold. In any event, the mice and rats grew more slowly and a higher percentage survived to the scheduled end of these long-term studies.

Speaking of regulators, feeding specifications in regulations promulgated under the federal AWA state (emphasis added as italics):

- "Dogs and cats must be fed *at least once each day*, except as otherwise might be required to provide adequate veterinary care" [Part 3, Subpart A, §3.9(a)];
- "Guinea pigs and hamsters shall be fed *each day* except as otherwise might be required to provide adequate veterinary care" [§3.29(a)];
- "Rabbits shall be fed at least once *each day* except as otherwise might be required to provide adequate veterinary care" [§3.54(a)];
- "Nonhuman primates must be fed *at least once each day* except as otherwise might be required to provide adequate veterinary care." [§3.82(b)].

Laboratory dogs are usually fed once or twice a day at discrete intervals, similar to how owners feed their pet dogs. And if and when the food bowl is emptied, we don't immediately add more. The same approach is conventionally performed for laboratory pigs and other large species. Laboratory cats are often provided dry food ad libitum because, just like pet cats, it's they, rather than we, that determine when they wish to eat. If canned cat food is provided, the practice is the same as for dogs, i.e., when the bowl is emptied, more canned food is not immediately offered.

By contrast, the usual practice for mice, rats, hamsters, guinea pigs, rabbits, and other smaller animals occasionally found in the vivarium, regardless of their coverage under the AWA, is to always fill up the food hopper attached to the cage with chow pellets, regardless of how much is left from the last feeding. This canon of keeping food hoppers always full for smaller species is enforced without hesitation by institutional and extramural regulators alike. Woe to a husbandry program if empty (or even near-empty) food bins are ever discovered in small animal housing rooms during a semi-annual IACUC or regulatory inspection or accreditation site visit. When detected by the in-house compliance assurance staff, the usual penalty is an internal citation (usually recorded only as "minor" rather than "significant" unless the animals involved are obviously underfed), with a corrective action plan assigned and tracked for completion. But what is the basis for this rigid orthodoxy?

The literal meaning of ad libitum from its Latin root is "at liberty" or "at one's pleasure."[7] This phrase somehow replaced the more appropriate "free-choice feeding," a practice conventionally applied to livestock nutrition (how and why such a switch occurred in a laboratory animal context could provide an interesting story, but will be saved for a later day). Since the AWA doesn't require such excess, what does the Guide say? The seventh edition (1996) stated that "Moderate restriction of calorie and protein intakes for clinical or husbandry reasons has been shown to increase longevity and decrease obesity, reproduction, and cancer rates in a number of species" and "Calorie restriction is an accepted practice for long-term housing of some species, such as some rodents and rabbits..." (page 40). The eighth edition (2011) similarly advises on page 67 that "Benefits of moderate caloric restriction in some species may include increased longevity and reproduction, and decreased obesity, cancer rates, and neurogenerative disorders" and "Caloric management ... can be achieved by reducing food intake or by stimulating exercise ...." So if we've been given allowance by the Guide for 20 years at least to avoid excessive feeding for better animal health, isn't it time to recalibrate expectations and mandates for ad libitum feeding?

One may argue that smaller animals have higher metabolic rates so they require more calories per unit body weight than larger animals over the same time period. That in turn is used to justify the practice that these calories must always be available. A logical riposte to this argument is that constant or even frequent feeding is unnatural for these smaller species, and their metabolism had naturally evolved in response to unreliable availability of foodstuffs rather than excessive consumption. Because sporadic access to food rather than non-stop caloric restriction is a more likely reality in the wild, it's reasonable to expect small species to actually be more rather than less inherently adapted to occasional interruptions in food supply. So it's no surprise that occasional fasting interspersed with ad libitum feeding has been shown to be just as beneficial to the health of rodents as prolonged caloric restriction, and possibly with similar outcomes for us.[8,9] If that's the case, then shouldn't we actually take all the food away from healthy, normal animals for short, defined periods of time?

One reason why everyone keeps hoppers filled is that it's just easy to do, and no hopper would ever run out of food so no animal could ever go hungry. But in cages with solid bottoms, food is almost always still available in crumbles that fall from the hopper and is mixed with the bedding on the cage floor. These food particles are safe to eat, especially if excreta are localized to a particular corner of the cage (and remember that rats and mice engage in coprophagy anyway, to obtain vitamins and other nutrients produced in the lower intestinal tract). Searching for food on the cage floor is also a more instinctive fit with a herbivore's natural foraging behavior, thereby providing another type of environmental enrichment. In fact, for rodents that are weak or in poor health, food is often provided on the cage floor as well as in the hopper so that these animals don't have to reach or climb to get to food (an IACUC exemption is usually indicated). None of this is to be misinterpreted as preferring the mixing of food in with soiled bedding. It's merely a reminder that when the hopper becomes empty, there's still food available somewhere as long as animals aren't on wire mesh flooring.

When food hoppers are exchanged to be washed, food left in the dirty hopper is sometimes discarded rather than transferred to a clean hopper, especially if the inhabitants of that cage are no longer alive. And new animals are provided with a full hopper of new food, regardless of how long they may be around. Thus, in addition to the potentially adverse effects on animal health from excessive eating, discarding uneaten food incurs considerable financial and environmental costs. My department recently conducted a small pilot study in which mouse cage hoppers were filled only 50% while monitoring animals in every cage to make sure they never ran out of food. The purpose of this exercise was to estimate how much food (and money) could be saved by simply providing less excess. For an average daily census of 8500 mouse cages, the difference between full and half-full food hoppers was 145 bags of mouse chow per month, equating to $35,000 in potential annual savings. And we were still providing much more than what was needed. Consider that the average adult mouse eats one chow pellet/day. For a standard-sized shoebox-shaped cage with a maximum of five adult mice, about 150 chow pellets per month should suffice under the convention of providing food all the time, especially if all five adult mice are around for the entire month. The standard food hopper for such a standard cage holds around 200 chow pellets when full, so filling the hopper only once monthly is still excessive under normal consumption patterns in a cage with the maximum-allowed occupancy. How low can one go without worrying about running out of food? According to the maximum needs calculated above, a 75% full hopper should suffice through two routine ventilated cage changes scheduled 2 weeks apart.

There are certainly caveats to embrace if one reduces excessive feeding and ventures closer to the possibility of animals not having food for short periods of time. For example, if the common laboratory mouse (*Mus musculus*) is without any food overnight, i.e., the time when it and other nocturnal species are most active, and if sufficient nesting material isn't available to stay warm below its thermoneutral zone, a temporary hypometabolic torpor-like state can result.[10] In mouse protocols that involve overnight fasting and barren cages, this adverse possibility should be taken into account. Another caveat involves the so-called "food grinders"—mice that chew but don't swallow food to such excess that food hoppers are quickly emptied while the cage floor becomes covered with the resultant dust. While there may be available nutrients in that dust, it's not an optimal situation because care providers often must add more pellets daily and must change soiled cages more frequently. Chewing devices and sunflower seeds have been shown to reduce excessive grinding, the latter remaining effective after their removal.[11]

The effects of excess intake in laboratory animals aren't limited to rodents. At a prior institution, we maintained over 100 macaques given nutritionally balanced biscuits ad libitum and obesity was occasionally observed in these animals on long-term experiments, a problem also noted elsewhere.[12] We calculated that monkeys were given an excess of daily biscuits equating to an extra 300 kcal more than what they needed. But we had maintained that practice so that every cage was never without food, day or night. Another nutritional reality that required adjustment was a consequence, in part, from a 1985 amendment to the AWA that required research institutions to provide environments promoting the psychological well-being of laboratory NHPs, including "varied food items."[13] One popular response to that mandate was to offer sweets and snack foods to monkeys since they often preferred those over the usual biscuits (no surprise), in an attempt to make these animals seem happier. But when junk food was provided to excess, a common result was heavier animals (also, no surprise). So in addition to the excessive number of regular biscuits, we were going further in the wrong direction. An animal nutrition scientist was engaged to revise our monkeys' diet so that fewer calories were consumed while ensuring daily nutritional requirements were met. A wide variety of foodstuffs was still included in the mix, but sweets and snack foods were replaced with a higher proportion of fresh fruits and vegetables, yogurt, nuts, etc. It was gratifying to see body weights respond accordingly.

Finally, let's return to the phenomenon of caloric restriction described at the beginning of this chapter. We know that occasionally going without food is more natural for most species of laboratory animals and reduced intake can extend their lifespans while avoiding or delaying common chronic ailments. This raises an interesting ethical question not recognized by current regulatory and accreditation mores, i.e., which is better animal welfare ethics: never hungry or healthier living? Anyone with a pet dog, especially if it's an active breed such as beagles, know that these animals will eat all the time if given the opportunity (one of my mentors, Leo Bustad, used to say that a dog is the opposite of a person in that if you arrive home late to prepare dinner, a dog is twice as happy to see you!). While the researcher's needs take priority, I wonder how IACUC and external oversight bodies as well as animal welfare protectionists would react if we intentionally limited food intake to below satiety levels? Other advantages include saving time and money, and reducing our environmental footprint. But the animal wouldn't be happy or appreciate that we're sparing it from premature aging and disease.

Chapter 2

# Smart Refrigerators and Dumb Cages

One of the hottest technology fields today is labeled the "Internet of Things" (IoT). It's defined as "the network of physical objects, devices, vehicles, buildings and other items which are embedded with electronics, software, sensors, and network connectivity, which enables these objects to collect and exchange data."[1] Everything from room thermostats, garage door openers, and medicine vials to entire homes, nuclear energy generation plants, and farm land are being monitored with devices that continuously gather data about their surroundings and transmit these data to other computers or digital dashboards for objective and prompted or automatic adjustments as the circumstances dictate. As a recent evidence on how hot this sector is, consider that Google recently acquired a "smart" home technology company, Nest, with not much more in their current product line than sophisticated thermostats, at a valuation of $3.2 billion in cash. Clearly, Google and many others expect the IoT platforms to catch on in a huge way, thereby justifying big bets such as this one.[2]

A simple example of technology-connected objects that caught my eye 15 years ago, long before the IoT moniker became popular, involves beer. Imagine you're enjoying your favorite brew with friends at a bar, and the mug in which that beer was served contains microchips that monitor fluid weight or level and communicate via radiofrequencies to a device your server is wearing. As you drink from the mug, it's constantly measuring how much beer is left and transmitting that information to the server's device, which is also monitoring every other customer's glass or mug at the same time. At some threshold predetermined either by the bar's owner or the brewer, a signal flashes on the device telling the server you may be ready for another round.[3] This avoids wasting the server's time by not coming over to your table before you may be ready for a refill and also doesn't bother you and your party if you want to be left alone. And if you've already had several beers, the device could also generate a yellow or red light telling the server that you may have had enough. The result? Definitely more precise service, likely more sales, and possibly lower labor costs because fewer servers may be needed.

Going back further in time, the common refrigerator has long been equipped with a variety of gizmos that automatically maintain temperature, avoid ice formation where it's not wanted, make ice (cubes) where it is, turn on the lights inside while the doors are open, and emit an irritating noise if the doors are left

Notes in the Category of C. https://doi.org/10.1016/B978-0-12-805070-5.00002-3

**9**

open too long, to mention a few. All of these features are good things since they help maintain the quality and safety of food and beverages that we keep inside refrigerators, unless we're not attentive and let things spoil.

What can consumers expect next? Manufacturers are developing even "smarter" refrigerators (i.e., those that could do even more things by themselves) that someday will make today's versions look like the old ice boxes that literally held a block of ice inside insulated walls to keep contents cool until the ice melted. It's reasonable to expect that refrigerators of the future will broadcast all their contents on a smart phone app and tell you what food or beverages are missing from your master stock list and that it's now time to buy more (or just automatically purchase replacements and schedule a delivery); tell you when specific groceries are near the end of their shelf life or are placed in the "wrong" location that will accelerate spoilage or absorb nearby and unwanted odors; and automatically adjust power consumption based on the kinds and quantities of various groceries being stored at the moment. All of this will be possible due to constant "cross-talk" between the refrigerator's sensors and those embedded in each bag, box, jar, and bottle of food and drink inside, appearing on your smartphone as a reader-friendly dashboard.

Why a focus on refrigerators? At its conceptually simplest, the refrigerator is a rigid and durable container that reliably protects fragile (perishable) contents from losing their value (financial or nutritional) before they're used. A rodent's cage also is a rigid and durable container that reliably protects fragile (living, perishable) contents from losing their value (financial or scientific) before they're used. Both types of containers require energy to maintain the prescribed environmental conditions for their respective contents[4] and both have been around for almost a century. What's the average dollar value of a refrigerator's contents these days based on what was paid at the supermarket? It's probably $200 or less on any given day, including expensive cuts of meat in the freezer,[5] and we've come to rely on these appliances protecting those "investments" without giving a second thought. Even the simple refrigerator door-ajar alarm has been around for more than 56 years.[6]

Contrast such handy features with today's rodent cage. We have sensors that monitor animal housing rooms and IVC racks for temperature, humidity, and air flow, all of which are nice to know but are of less use when the rodent's immediate environment inside a micro-isolation cage with a snug lid can differ from what's going on in the room. We have sensors that monitor automatic watering systems to identify which rack may be drawing more water than normal but can't localize which specific cage may be flooding or detect individual water valves that are stuck in the off position that don't provide *any* drinking water to animals. Instead, we rely on a person, usually an animal care technician, to visually inspect each cage in order to comply with the mandate that every animal be observed at least once daily. This visual assessment is intended to see if the animals are okay, if there's enough food, if the food isn't spoiled, if the water source (automatic or bottles) is working, and whether the cage needs changing ahead of schedule.

But in vivaria that contain thousands or tens of thousands of rodent cages, it's not practical to pull out every cage to check thoroughly on each mouse or rat (it's also a bad idea to do that anyway because a daily pull is physically and behaviorally even more disturbing to the animals inside, creating literally seismic activity on a rodent scale). Accordingly, most programs rely on an initial visual inspection without touching or handling the cage, often using a small flashlight for better observation. But if the animals are hiding in the back of the cage because they're naturally nocturnal and prefer darkness, or staying cuddled up in their nests to stay warm because the room temperature is below their thermoneutral zone and too cold for their liking, the technician has to either pull out the cage or make an educated guess and move to the next one of the hundreds assigned to his or her daily routine. That's it; no fancy sensors, no intra-cage monitors, no dashboards, and no alarms when elements inside the animal's home are out of whack.

If the rodent cage is so much more mechanically primitive than the refrigerator, the cage contents must certainly represent a much smaller monetary investment than groceries, right? Sadly, it's not even close. Although there are no statistics I could find, a preliminary approximation comes in at a total of $7.87/mouse cage/day in sunk research expenses, before monetizing any subsequent value generated by that research.[7] Apply that total to 120 mouse cages on a double-sided IVC rack with a capacity of 140–160 cages, and the total invested dollars at risk for cage or rack malfunctions that could torpedo that investment on any given day is around $1100 (or slightly more than five times my estimated value of the groceries in your refrigerator). Or put another way, for a representative barrier rodent facility maintaining an average daily census of 8000 cages, the total sunk research costs could be almost $63,000 on any given day or almost $23 million for an entire year. Yet we remain content with once daily, brief visual glances of animals often difficult to see and rodent houses devoid of automatic and continuous monitoring capabilities to protect that investment as well as ensure animal welfare.

Let's say we had smart rodent cages. What would they do? As with any commercial product development process, one should start with customer needs. All technology added to today's rodent cage should be (1) durable enough to withstand multiple sanitizing and sterilizing cycles (or be easily detached to avoid being damaged while the cage or rack is being moved and cleaned); (2) fail-safe (if it's not reliable, including avoiding false alarms, then no one will buy it); (3) no or low maintenance, just like the cage it supports; (4) simple to program, use, and interpret so that non-technical personnel such as night watch crews can respond to alarms; (5) able to retrofit to the already purchased cages or racks so that one's existing capital inventory doesn't need to be replaced; and (6) dirt cheap on a per cage basis. That last item is especially important for many reasons, including but not limited to the fact that the parent institution that pays for the equipment and subsidizes the animal care budget isn't the ultimate beneficiary of any smart cage enhancements. That, of course, would be the researcher who usually has much less say over capital investments for the building in which she or he works. So a cheaper acquisition price lowers the resistance of the institutional buyer to making the investment on behalf of the end user.

Given these basic requirements, what features would one like a smart cage to have? What elements would be helpful, if not vital (literally, if you're the mouse), to monitor and issue an alarm whenever these elements begin to drift dangerously outside the tolerance range? The first commercially available mouse cage sensor package was recently released and targeted moisture, soiled bedding, animal activity, and levels of food and bottled water to safeguard a comfortable environment for the animal and to detect flooded cages, while indicating the current cage location.[8] One hopes that other real-time, intra-cage monitoring technology platforms will have been introduced to the market by the time this is published, maybe with additional or different features.

Monitoring animal activity inside the cage is intriguing because it could be an effective animal welfare monitor, much more informative than any approach currently used. Because mice are nocturnal, they're not doing much when husbandry personnel are around during the day. Accordingly, it's sometimes hard to tell if mice are sleeping or sick, not to mention if they're getting better or worse, during that brief daily visual assessment. By contrast, at night each cage might as well be a venue for a track meet, with mice busy scurrying around, having a meal or two, digging through their bedding, rearranging nesting material, and getting a serious aerobic fitness workout if there happens to be an exercise wheel handy.

Imagine all of this activity establishing a baseline for comparative purposes, possibly adjusted by the number of animals per cage, as well as their age, sex, and genotype. The activity readings outside an established normal range could trigger an alarm that requires timely, if not immediate, inspection of the animals. Too much activity? Perhaps it's a cage of adult male mice from different litters that are fighting and need to be separated before serious fight wounds are inflicted. Alternatively, what if the mice are supposed to be hyperactive because of their phenotype or as an intended effect of the drug they had been given? Then an excessive activity reading could provide valuable research data and spare some space in the cage by avoiding the need for an exercise wheel also to measure that activity. Too little activity? If the mice recently had surgery or are being treated with analgesics for another reason, maybe the mice are reluctant to move because their pain medications aren't working. That could be a signal to evaluate the animals, regardless of how they look during the daytime, and possibly adjust their drugs or dosing regimen. What if the mice are on a protocol where they're anticipated to get sick, moribund, or die? Employing a constant measure of activity could more accurately track their decline and prompt one to remove the harmful stimulus or euthanize the animals before they suffer further, if that's compatible with the scientific aims of the experiment.

What other tracked elements for mouse cages could have just as large an impact on laboratory animal care and use? Borrowing from a real estate broker's standard pitch: "location, location, location." One of the biggest wastes of time in the vivarium today is searching for a specific cage, even if one has the correct room and rack. Sure, there are cage cards that identify individual cages by the assigned IACUC protocol number and the name of the principal investigator. But that still requires one to be knowledgeable about whatever specific cage card system has been implemented locally. And there are usually lots of different kinds of cage cards, varying not only by text but often also by color, size, and shape. If the veterinary technician or veterinarian is notified that a mouse needs a closer look, there still may be other cages with an identical rainbow of cards adjacent to the cage of interest. So you have to read the (usually) fine print on sometimes more than one card on a given cage to make sure you've got the right one. For researchers, it's often worse. They'll get a notice that their order of 30 new mice has arrived and has just been housed in Room X, on Rack Y, Side Z. If these new cages are the only ones on that side of that rack, then finding them is easy. Usually that's not the case—the old and new cages may be mixed together, as well as intermingled with other researchers' cages. Making matters worse, today's array of cages on a rack or in a room is often transitory, especially in academia, where the cage location is not precisely controlled as is often the case in the industry. A mouse may be transferred to another cage in another slot on the same or different rack by laboratory members in order to group animals conveniently closer in advance of the next experiment, or because these mice were breeding or fighting and needed to be separated, or because a litter of pups needed to be weaned by sex into new cages and no adjacent slots were available. I haven't figured out a reasonable way to quantify the time everyone spends looking for particular cages but am sure that it's a huge number when calculated over an entire year for an institution, and then extrapolated for the entire research community.

Contrast that with driving into parking lots at shopping malls and airports. Upon approaching the entrance, a big display board indicates which floors still have how many empty spaces, or in the case of rental car agencies, where my reserved car is located. I don't have to waste time driving around in search of a spot to park that doesn't exist on that particular floor. And after I land from my return flight and pay for parking, my exit card magically reminds me where my car is located. Wouldn't it be nice if details about one's cage location were instantly accessible and always current? Someday, there will be smart phone apps that provide such information, password-restricted such that only those persons who are on a given protocol will be able to find where their cages are at any time from anywhere they happen to be when querying the app, whereas other laboratories would be password-restricted to locate and track only *their* cages. The vivarium staff would be able to track everyone's cages, of course, and having a digital record of the cage location and movement would enable retrospective analyses in case one wanted to know where specific cages had been earlier if there's a localized infection outbreak or environmental systems failure. The display of this app would be a dashboard that depends on hardware that would always match each cage to its current slot, an automated "handshake" of sorts between cage and rack. If a cage is moved, a new location would be signaled upon sliding that cage into its new slot and the old address would end. Over the past 10 years, digital tracking technologies have become commonplace for taking cage census for per diem billing. But these technologies, such as bar codes and radio frequency identification (RFID), can so far only localize a cage to a room or a rack and not provide the desired address details.

Another element that's often mentioned whenever smart cages are discussed is intra-cage ammonia. It would be nice to monitor ammonia from excreta so that its buildup isn't excessive and irritating to the eyes and mucous membranes of mice or other inhabitant species. When ammonia levels exceed an established maximum, a signal could be issued that it's time to change that cage. But there's no evidence that excessive ammonia is a problem in IVCs that are properly managed. The number of air changes per hour inside each cage is usually sufficient to avoid this problem before the cage needs changing based on the conventional engineering standard of every 2 weeks or by common visual criteria employed for "spot changes," i.e., when the cage looks excessively soiled. Perhaps there's an argument to be made for ammonia sensors in static cages where air circulation is much less than in IVCs. However, static cages represent a small and continuously shrinking slice of the market so development costs of an ammonia sensor for a smart static cage won't likely be recovered by subsequent sales. In either event, any chemical sensor is likely to violate at least one of the basic industry requirements listed above (too expensive, too fragile, etc.).

Looking further into the future, one can envision other parameters for continuous measurement that are more mouse-centric than the simple elements that we can measure cheaply and reliably today. For example, a couple of years ago, Harvard scientists reported that they were able to transmit smells between Paris and New York over a smartphone app.[9] This amazing feat shared the fragrance of French perfume and the aroma of a tasty midtown Manhattan breakfast between the two parties. We know that smell is the dominant sense in mice and many other laboratory animals. If we could identify and isolate the smells that are most important to these species, such as sex pheromones and other behavioral cues, it stands to reason that chemical sensors tracking these smells and alerting us whenever they're detected or outside of an established range could tell us more about what the occupants of these cages are feeling and perhaps optimize their care even further. Analogous arguments could apply for light sensors calibrated to the ultraviolet spectrum where mice can see but we can't or for microphones designed to pick up ultrasonic sounds where rodent hearing is focused so that we could monitor the chatter amongst cage mates and intervene sooner if indicated, either for managing breeding[10] or to monitor the efficacy of analgesics in managing post-operative pain.[11] What comparable positive and negative stimuli of a visual, auditory, or olfactory nature and beyond our range of detection could prove indispensable to monitor other species for better welfare and better research?

One doesn't have to end this wish list for just stationary cages. What about "cages" used for transport, otherwise known as animal shipping containers? Currently, we don't measure or transmit environmental conditions or other animal welfare parameters inside these containers during transit. But it would be good to know if animals are too warm or too cold and by how much while they're on their way to their destination. We know that adverse transport exposures can affect animals' physiology for extended periods of time, but these data have relied on the animals themselves being instrumented with sensors that compare vital signs, circulating glucocorticoid levels, and other stressor indicators before packing and after arrival.[12] On a much simpler and less invasive level, some rodent breeders are beginning to include digital environmental monitoring devices inside shipping crates. These types of devices are commonly found in what's known as the cold (supply) chain industry for monitoring conditions under which food and other perishables are shipped.[13] Recently, they've been exploited by rodent breeders for improved animal transport container design as well as quality control of the living cargo.[14] In addition to ambient temperature and relative humidity, it would be useful to know the spectrum, intensity, and duration of other environmental parameters, such as light, carbon monoxide (from vehicle exhaust), sound, mechanical vibrations, or jarring and tilting to which the animals are exposed during transit. The duration of actual motion in transit versus sitting still, as well as the route and current location, could also be informative with respect to the effects on animals.

If these in-transit monitoring devices are already proven to help protect precious cargo in other industries, why aren't they already an established component of shipping laboratory animals across the globe, country, or even towns? If cost is a concern (as opposed to an excuse), then should not suppliers of laboratory animals offer intra-transit monitoring at a reasonable markup and let customers decide if they want to include it in the shipment? And if they did, what would we do with that information? Certainly, we are obligated to our customers, the researchers, to provide them with as much information as possible about the quality of animals that we receive and care for. Data about intra-transport conditions could lead to better recovery and acclimation procedures post-arrival, more informed instructions to the vendor or shipping handler, and improved containers, as well as grounds for switching shippers or vendors.

But these possibilities generate even more questions. On the practical side, what is the optimal sensor location? If inside the container, how small and sequestered from the animals must it be? If outside the container, how representative of the animals' experience inside are the data it captures? What about sensor durability, reliability, and safety to the animals? On the marketing side, how much more would you pay for such animal transit data? How often should transportation conditions be monitored—every shipment, occasional spot checks, or perhaps seasonally? Would you be willing to pay a premium for guaranteed transit conditions (or free replacement animals)? Or would you offer to pay for inserting your own data logger in a vendor's container or vehicle and would the vendor be willing to cooperate?

If and when smart cages ever become common, they in turn will change how routine animal care is performed. Just like the smart beer mug frees the server from making unnecessary trips to your table to see if you could use more beer, the smart cage will free technicians from checking on it in the absence of an alert that there's too much or too little activity, that it's too moist, too cold, etc. The standard and not wholly reliable practice of observing every animal at least once daily to see if it's okay (often a subjective judgment) will be replaced by constant and quantified scrutinizing of the contents of that cage, both live and inanimate. Only when a predetermined tolerance range is violated and a warning issued, will a cage actually need to be inspected. And the inspecting technician, as well as everyone else connected to the monitoring system, will have details on which element is the problem before ever arriving at the cage. Even better, these tolerance ranges will be adjustable so the mice with special characteristics or needs can be monitored with more precision and given faster service than before. Take, for example, mice that are expected to be hyperactive. The activity sensors in these cages would have a different tolerance range of readings from cages housing regular mice so that alerts would correspond to deviations from a different "normal" range for them.

If remotely monitored cages don't have to be observed visually at least once daily by animal care technicians, does that automatically mean we need fewer animal care technicians? Probably not. That's because, with cages always being monitored, alarms could be activated at any time, including and especially in the middle of the night when mice and other nocturnal species are most active. Depending on the nature and severity of the alarm, an assessment and response may be required sooner rather than during the next workday when that cage normally would have been visually inspected and the problem hopefully detected. In other words, we couldn't any longer plead temporary ignorance of a problem and an animal at risk or in grave danger. If an issue in a cage demands an immediate response, then we would have to respond immediately in order to avoid or minimize damage to the animal and the experiment. And because of the variety and specificity of elements potentially going into alarm, the approach common today of relying on the overnight physical plant personnel to investigate and resolve a cage malfunction likely wouldn't be sufficient. In addition to the usual array of HVAC and automatic watering glitches, responding to a wide spectrum of biological conditions in each cage will likely be more than what physical plant employees are able to handle.

Instead, it's reasonable to imagine that vivaria will have to be staffed around the clock with animal care technicians empowered to intervene knowledgeably and promptly when cage sensor alarms go off. Waiting until the next morning will no longer be an option. How many technicians will be needed per 1000 cages? Will that offset labor efficiencies gained from automated, remote cage monitoring? Would on-site supervision of employees during the wee hours of the morning be necessary, or will facility supervisors have to be on call 24/7?[15] Either way, will more facility supervisors need to be hired to ensure monitors are working as expected and technicians are responding as required? As with many technological advances, it's often the human considerations that are not included in the predicted outcome.

That introduces another anticipated lesson that smart cage companies will need to learn, likely more than once. Many durable goods manufacturers are scrambling to enable their machines to communicate with each other and with customers in this emerging IoT universe. As new smart products are rolled out, it's becoming obvious that the smart product's value is increasingly dependent on the IoT software and the data it transmits rather than the machine itself. This in turn means that manufacturing companies will need to pivot and recognize they're now in the information business much more than they're in the equipment business.[16] Buyers of smart machines will expect their products to come with user-friendly dashboards, preferably on a smart phone app, which aggregate the torrent of data from these machines so that the current states and outliers are easy to understand and manage.

This leads us to the discipline of analytics, another emerging tool becoming just as integral to businesses as IoT objects.[17] Initially, analytics were employed to identify retail purchasing trends and other mass market data sets. But this field is expected to quickly become predictive for more successful positioning of new products, prescriptive for semi-personalized health care decision making, and eventually automated so that IoT-embedded machines become self-adjusting in response to changing inputs.[18] But it won't stop there. Is a given smart technology platform expandable to other inputs peculiar to a given customer? Will that require an open rather than a closed source software code? All these analytics questions and amazing possibilities also directly apply to the laboratory animal care industry and eventual roll out of smart cages. Buyers shouldn't and won't be satisfied with simple alarms only when cages are out of pre-set tolerance ranges. Instead, the laboratory animal care staff will want to know which cages of what kinds of mice at what times of the year in which facilities have been problematic so that they can avoid future problems on a programmatic rather than ad hoc basis. Investigators will want to integrate their research data capture needs with smart cage customized to the intellectual pursuits of each laboratory. This isn't meant to imply that smart cage vendors have to become overnight experts in Big Data. But in order to succeed, they will have to make some strategic and expensive decisions way beyond merely figuring out how to incorporate sensor hardware into plastic boxes.

Let's say that all of the above can be figured out and leaves us with a positive case to make for taking the plunge and incorporating cage sensor technology in everyday vivarium operations. Certainly the magnitude of at-risk research costs already invested in each cage should demand at least a serious look at the capital and operational costs involved. And the core technology has been around for years and embraced by other industries for continuously monitoring fragile systems, so it's not like anything new has to be invented first. What's holding back our field from demanding such features this minute, if not already using them as a matter of habit?

One barrier to rapid adoption is that it's simply too early. Not enough products are available and not enough experience has been gained and shared from the single technology package that's currently being marketed. Because this is so new, at least to our community, more time is inevitably needed to sort what works well from what works less well or doesn't work at all. As another example, I'm reminded of the interest we had in RFID technology for cage census taking around 12 years ago when I was directing the program at MGH. RFID had already become a common tool in many other industries, from consumer product retailers (Wal-Mart and Proctor & Gamble) for supply chain management to pharmaceutical firms (Johnson & Johnson) for brand security. There was a great television commercial at that time that started with a suspicious looking guy nervously avoiding the checkout counter while exiting a supermarket, at which point he was immediately approached by a security guard. The viewer was supposed to think this guy was going to be busted for shoplifting. Instead, the guard merely said, "Sir, I think you forgot your receipt" and handed him a long piece of paper. All of the goods this guy was carrying had been automatically scanned by an RFID station and charged to his credit card while he headed for the door, without requiring any cashier to tally his purchases and collect payment.

At that time, we were attracted to RFID technology because we were looking to replace our bar code identification and counting program for tracking over 13,000 mouse cages. We had just purchased the bar code system in 2003, tagging every cage with a unique bar code that was scanned by hand regularly. This replaced the prior manual "system" of a pencil and clipboard for counting cages, and bar coding enabled better accuracy, faster turnaround times,[19] and a digital census record for easier audits and occupancy management decisions. Nevertheless, bar codes required paying animal care technicians to scan each cage card every time that we counted that cage. The labor needed to perform this task alone grew along with the expanding number of mouse and rat cages in our program, eventually costing us $200,000 per year in technician wages and fringe benefits. RFID was an attractive alternative because cages with RFID tags would be counted automatically by ceiling-mounted receivers as often as we wanted, with no labor involved. Unfortunately, no RFID systems to automatically count cages were commercially available at that time so we had to play around with simple mockups[20] until vendors rolled out their versions in 2008 for a closer look.

Another hurdle that smart cages have to overcome is a consequence of the administrative disconnect, mentioned earlier in this chapter, between the capital buyer (the institution) and the ultimate beneficiary (the researcher), especially if institutions have to procure caging equipment from general funds that have so many other competing claims for money. However, if those who underwrite sponsored research in academia or perform it in their own laboratories better understood how automatic remote monitoring would better protect their intra-cage investments, perhaps some of their funding would be reserved for smart cage acquisitions.

Our earlier approach to commercial RFID census platforms may offer lessons to be learned in this regard, too. Rather than writing a big check to buy the entire package and then hoping it worked as advertised, we rented a smaller version from one vendor for 6 months, just to see how it would perform in our vivarium. This approach minimized our capital risk while allowing us to see how easy the RFID census software could communicate with our per-diem billing software. And it was a good thing that we took the test drive approach—most of the walls in our animal facilities were made of sheetrock rather than masonry, which unfortunately didn't block RFID radio waves from traveling through the walls and counting cages in adjacent rooms. As a result, cages could be counted more than once with every RFID transmission burst and not be reliably localized to their actual room of occupancy. So we backed off at that time and stayed with bar codes. The same piloting strategy could also work for smart cages, i.e., try before you buy.[21]

Finally, widespread adoption of smart cages will also have to wait because of our field's resistance to change. Laboratory animal care has been driven by convention since its inception. This approach was understandable during the early days when little was known about keeping animals in laboratories. But we've institutionalized the ethic of engineering standards so deeply in our collective psyche and continue to reprimand those who don't follow every word of the Guide verbatim, even though it was always meant to be a small-case "guide." The result, locally, is program leadership that avoids risk in order to avoid blame, and, nationally and beyond, deep skepticism over anything new. Until these attitudes are relaxed or replaced by the next generation of program leaders, refrigerators will continue to be smarter than cages.

# Chapter 3

# Thermoneutral Zones, We Hardly Knew Ye

It's a fallacy that everyone likes surprises. For example, some surprises are irritating if they keep happening but aren't supposed to (think Phil Connors, the lead character played by Bill Murray, in the 1993 movie "Groundhog Day"). Like that storyline, the surprises about which I'm thinking were not expected to be repeated, making them all the more exasperating. Let me explain.

Whenever surprises occur in science, it's often a big news event. That's because researchers are usually characterized as plodding along clearly defined paths, taking the next incremental step in discovery dictated by the most recent findings reported by someone else's laboratory if not their own. And that's often how science progresses, one baby step at a time, and not always in a forward direction. But several years ago, news splashed about some published results involving laboratory mice experimentally given cancer. The paper of interest described how some mice housed at an ambient temperature commonly maintained in vivaria (around 22–23°C) had different responses to tumor growth and anti-tumor drugs compared with mice administered the same kinds of tumors but housed in warmer environments (30–31°C). In fact, housing mice in warmer cages resulted in a "striking reduction in tumor formation, growth rate and metastasis."[1] The popular science press latched on to this finding and the implication that most, if not all, prior research data extracted from mice housed at lower temperatures could be suspect. Headlines such as "Brrrr-ying the Results" and "Lab Mice Are Freezing Their Asses Off"[2] generated lots of buzz and likely caused angst for some researchers.

Laboratory animal veterinarians were scratching their heads about this so-called revelation, wondering why this was so unexpected. It's common knowledge in biology that the smaller a homeothermic animal is, the higher its basal metabolic rate (BMR) in general terms. This relationship was first quantified by the insightful physiologist, Max Kleiber, in which an animal's metabolic rate, stated in kcal/day, $= 70M^{0.75}$ with M being the animal's body mass in kilograms.[3] Since then, many studies have delved deeper into the inverse relationship between the size of a bird or mammal and its BMR, including how ambient temperature can affect the BMR as the animal tries to cope in different environmental temperatures. This led to the concept of the thermoneutral zone (TNZ), defined by Gordon in his excellent review of the topic as "a range of

Notes in the Category of C. https://doi.org/10.1016/B978-0-12-805070-5.00003-5

ambient temperatures where metabolic rate is at basal or resting levels," such that "within this range of temperatures,... a stable core temperature is achieved with adjustments in insulation, posture, and skin blood flow and there are no active mechanisms of evaporative water loss."[4] Lots of published data are out there that highlight the metabolic price that a laboratory rat or mouse must pay to stay warm at a temperature we prefer. One paper from 2004 showed a highly linear correlation between the changes in ambient temperature over a range of 18–30°C and changes in the mean arterial pressure, mean heart rate, and mean blood pulse pressure of laboratory rats and mice (and because mice are the smaller of the two species, some of their physiological adjustments were predictably more sensitive to temperature changes than those of rats).[5] One paper published 60 years (!) before this one documented the effect of varying ambient temperature on animal research data.[6] Therefore, it's no wonder, and should have been no surprise given today's ubiquitous online search engines, that tumor growth, drug metabolism, immune responses, and other physiological phenomena could be influenced in situ when an animal lives for any extended time outside its particular TNZ.

In the earliest days of keeping rodents in laboratories for experimentation, none of this was known or appreciated. Small wooden boxes fitted with wire mesh floors and lids, supplemented with a water bottle and perhaps a food hopper were deemed sufficient for adequate housing. Fast forward to the industrialization of animal research, beginning in the 1960s, when scale and efficiency were dominant considerations. Cages retained wire mesh flooring, so all the excreta could fall through the holes in the mesh for more efficient cleaning underneath. In addition cages were made of galvanized steel, eventually to be replaced by stainless steel, so body heat was quickly conducted away while the poor animal was left trying to stay warm by generating more body heat. Meanwhile, the recommended temperatures in the Guide and other authoritative standards for animal housing rooms remained within the human's rather than the rodent's TNZ.

Even with the advent of transparent plastic shoebox cages with solid bottoms in the 1980s, it was believed prudent for years to provide only a minimal amount of bedding in cages so that animals could be more easily observed for better disease detection; if too much bedding was dispensed in each cage, a nocturnal animal would hide in it to sleep while the lights were on. If that animal happened to be sick, be lame, have a skin laceration, etc., it would be more difficult, if not impossible, to see such a problem until the animal was in much worse shape. Only within the past 15 years or so have we returned nesting substrate to the mouse cage, primarily to give the animal something merely to do in order to "enrich" their behavior rather than for thermoneutrality, i.e., for the purpose of *making* nests rather than huddling *in* them. Only more recently has the heat insulation value of mouse nests become appreciated as an "added" benefit for better thermoregulation. To that end, the current edition of the Guide advises on page 43 that "...animals should be provided with adequate resources for thermoregulation (nesting material, shelter) to avoid cold stress." This recommendation is backed up with studies that show mice conserve body heat more easily and appear more comfortable when provided sufficient nesting material.[7] Furthermore, because nest making is so deeply engrained in normal mouse behavior, the failure of mice to make nests soon after a substrate is introduced in the cage can be a handy indicator that such mice are feeling sick or weak.[8]

Now that we've defined the problem (again) and instituted some improvements regardless of their original rationales, how do we make it even better? And I define "better" in two ways. First, how do laboratory animal experts make sure that scientists don't forget (again) about the reality of TNZs and any potential adverse impact on their research? Second, what additional improvements are practical for today's housing systems for laboratory rodents and other similarly impacted species to accommodate their TNZs?

To the first question, several programmatic solutions come to mind. With the advent of electronic IACUC protocol forms, why not simply include a question in the form that asks the principal investigator if the intra-cage temperature is an experimental concern—yes or no? If that's too subtle, how about inserting a statement that "housing smaller animals with higher metabolic rates in rooms maintained at temperatures below their TNZ may influence experimental results," followed by a box that one has to check to acknowledge that the statement has been read? Such cues could be augmented by additional reminders during new investigator orientation and in refresher training on responsible conduct of research, as well as by signage in the vivarium such as "Do you know if your animals are comfortably warm today?" But merely highlighting the potential risk is not enough. There need to be options available for investigators to employ in order to ensure that TNZ discrepancy is not a problem. Which brings us to the second question, i.e., what alternatives are available for getting closer to rodent thermoneutrality that are both effective and practical?

In the 1940s, some laboratory mouse colonies were maintained at 80°F (27°C).[9] Even today, a prominent gnotobiotic mouse facility with which I'm familiar keeps its mouse rooms around 28°C for the explicit purpose of animal thermoneutrality while it continues to perform groundbreaking research in cancer biology. But for the rest of us, raising room temperatures wherever rodents are housed is neither an attractive nor affordable strategy in an era of energy conservation and tightening research budgets. And workers, especially those in temperate climates, wouldn't be happy about hotter working conditions.

[That raises an interesting question in the face of all-too-predominant wealthy nation chauvinism. What about vivaria in subtropical and tropical regions? Do they really have to adhere to husbandry guidelines established by developed countries in temperate or subarctic climes that can more easily afford the energy costs of maintaining narrow bands of environmental parameters? What would happen to laboratory mice if they were housed at warmer temperatures by institutions with perhaps less financial resources to pay for expensive air cooling? The answer seems to be that, at least for mice, they would do just fine. One study showed that mice maintained in IVCs at 28°C instead of 22°C showed no adverse effects on reproductive performance.[10] So depending on the animal species involved, we should cut colleagues in warmer climates some slack and not insist that what's conventional for humans at temperate latitudes should be followed lockstep elsewhere.]

As an alternative to increasing the temperature of the room or an entire facility, can we miniaturize the space in which there needs to be more heat? One approach is to lower the velocity of air in the cage so that the heat generated by the animals is not immediately evacuated. Some IVCs create a colder intra-cage environment if the air exchange rate is high enough.[11] By contrast, so-called "static" cages (those without an active air supply and exhaust) avoid this problem because passive air exchange through the mesh filter top is sufficiently slow. But even if the cage is actively rather than passively ventilated, a slower air velocity still allows some heat generated by the cage's inhabitants to warm the air inside that cage while being removed less quickly. Whether static cages or low-velocity IVCs are used, I have observed that the effect of lower air exchange rates is readily apparent in mouse behavior. Rather than being huddled together in one corner to stay warm, mice in such cages are dispersed and more active, even during daylight hours. However, there's no free lunch, and the trade-off with static or low-velocity IVCs is usually more frequent cage changing and processing. That's because fewer air exchanges per hour means more ammonia buildup inside the cage from ever accumulating excreta. That in turn results in higher labor, materials, water, and energy costs as well as disturbing the mice more often.

Are there ways to keep rodents comfortable in a cost-effective manner? One could heat air either supplied to or already inside the cage. This would avoid heating the entire room without sacrificing the benefits of a higher air exchange rate to avoid ammonia buildup inside the cage. We have explored a way to warm the local air supply to one row of ventilated cages at a time by inserting a simple electrical heating element into the air supply plenum; preliminary findings showed that temperatures inside cages supplied by that plenum remain warmer, whereas the cages in adjacent rows were unaffected.[12] And at least one vendor is offering a means to heat air already present inside the cage.[13]

Despite all the above, let's take a step back and reconsider how much we really know about the trade-offs between thermoneutrality, adapting to slightly colder temperatures via compensatory behavioral and physiological responses, and potential influences on research data. Some argue that as long as laboratory mice are group housed or provided adequate nesting material for insulation, the resultant gap between their TNZ and the ambient temperature in which they are conventionally housed is no different than us wearing clothing in the ambient temperatures that we normally work and live in.[14] By turning up the heat for these animals, we risk creating an environment that less accurately mirrors human metabolism and may skew data in another way.

Going forward, the least we can do is to remind current scientists and educate the next generation about this basic physiological property of laboratory animals so that they can incorporate these variables, if important, into their experiments. An even better solution is to design and offer micro-climate strategies for captive animals that optimize their comfort without threatening our own when indicated. And ideally, such strategies would be localized so that adjacent cages or rows of cages could have air that is conditioned to different specifications. A combination of practical insulating substrates inside cages and mechanized strategies to supply these cages with temperature-appropriate fresh air can go a long way in avoiding future "surprises" in how straying from an animal's TNZ may unintentionally influence research findings.

Chapter 4

# Waiting for SPF Zebrafish

To say zebrafish have become popular laboratory animals in discovery research, environmental toxicology, and other life science fields is an understatement. Initially taken from irrigation ditches and other semi-static water environments in south central Asia, the laboratory zebrafish was popularized as a research animal in early 1980s by George Streisinger and his colleagues at the University of Oregon. By exposing zebrafish embryos to radiation and mutagenic chemicals and seeing how their organs developed as the mutant embryos matured, combined with cross-breeding and genetic linkage analyses, these and other scientists were able to trace how vertebrate hearts, alimentary tracts, etc. formed from progenitor cells in a normal or altered progression.[1] This foundation in turn encouraged further investigation to understand how cells would develop abnormally or under specific disease conditions.

Today, nearly every major biomedical research institution houses at least one zebrafish colony, usually numbering hundreds or thousands of tanks. These colonies could be mistaken for pet stores, with lots of little glimmering fish in clear plastic containers on rows upon rows of shelves or racks, until one notices the barren concrete or epoxy floors, the absence of windows, and massive water quality systems (as well as the absence in the tanks of the obligatory miniature shipwrecks or mermaids). As evidence of the prominence of zebrafish in the laboratory animal menagerie today, over 19,000 scientific papers and books have been published over the past 10 years that included the words "zebrafish" or "zebra fish."[2]

While zebrafish research has moved forward, the same wasn't always the case for zebrafish husbandry and medicine. For many years, details about their care weren't considered important enough to merit a formalized dialogue amongst those responsible for maintaining laboratory zebrafish in their complicated housing systems. Years ago, I reached out to Claudia Harper, an accomplished laboratory animal veterinarian with a strong background in fish medicine, to see if she was interested in joining my department at MGH, in part to establish a veterinary resource for the many zebrafish researchers in the Boston area. She agreed to come on board and we began looking at unmet needs in that space. Perhaps establishing a regional diagnostic laboratory to detect infectious diseases would be welcome or maybe offering professional consultation for laboratory fish that fail to thrive, regardless of the reason, would get some traction locally.

Notes in the Category of C. https://doi.org/10.1016/B978-0-12-805070-5.00004-7

One day, we were sitting in my office, lamenting the disconnect between the abundance of national and international societies devoted to zebrafish biology and research, and how none of those organizations reserved space or time for zebrafish care. As a result of this bias towards research models versus routine care, there was much less information available about husbandry standards and strategies despite the fact that how the fish are raised and maintained can influence the experimental results they generate. Claudia said she knew many knowledgeable zebrafish colony managers in the Boston area who were tackling zebrafish husbandry issues on their own but had no easy means to compare notes with their peers. I suggested we start a regional association to bring them together to share problems and solutions and even possibly organize multi-site "trials" to evaluate the relative merits of various approaches to water quality, fish nutrition, and infection control, to name a few. Claudia agreed to contact a few key players to spread the word, and our department would host the first few meetings, throwing in free lunches to encourage attendance. Thus was born the New England Zebrafish Husbandry Association. A group of 17, compromising academic zebrafish colony managers, commercial vendors, and me as the tolerated outsider, gathered on December 1, 2005. At that inaugural meeting, we established initial organizational objectives, prioritized topics of greatest interest, and considered various tactics to spread the word to colleagues in Massachusetts and beyond. The meeting ended with agreement on a mission statement, "To promote and develop zebrafish husbandry standards through education, collaboration, and publication," that remains largely intact to this day.[3]

One important difference between laboratory zebrafish and laboratory rats and mice today is their respective natural infection status. As recently as the 1980s, rodent colonies were rampant with pathogenic microbes. The good news is that these colonies are now essentially SPF. This was accomplished over several decades by a combination of (1) research to identify various murine viruses, bacteria, fungi, and parasites and to understand how they may impact experimental data; (2) standardization of diagnostic assays and colony health surveillance practices; (3) investment in housing technology that simplified isolation of animals from each other to prevent infection between neighboring cages. Cleaning up the colonies also required convincing scientists that it was in their best interest to upgrade their experiments with "better" SPF animals; this was especially challenging when the animals looked healthy but still harbored bugs that could alter research results. A key enabler for all of this to happen was the advent of the so-called micro-isolation cage and accompanying laminar flow transfer hood or changing station that provided a purified air environment within which one could open the cage and protect its inhabitants from whatever viruses or bacteria were wafting through the air outside the hood's chamber. This combination and practice effectively collapsed the protective barrier to just the cage itself rather than the entire room or even the entire vivarium. With these rodent housing components in hand, airborne or fomite infections were prevented from jumping from cage to cage, without personnel having to don personal protective equipment or shower in or out of facilities to avoid spreading unwanted bugs around.[4]

By contrast, the infectious microbe status of today's laboratory zebrafish is less well defined and not as tightly controlled. Maybe because zebrafish are cheaper than mice to acquire and maintain (i.e., sick or dead individuals cost less to replace) and are possibly more hardy, there hasn't been the same pressure by or on zebrafish users yet to rid their zebrafish colonies of infection even though strategies have been devised to avoid serious outbreaks and to clean up colonies afterward.[5]

According to Chris Lawrence, who manages the Aquatic Resources Program at Boston Children's Hospital and knows as much as anyone on the planet about maintaining laboratory zebrafish, a sustained and effective campaign for SPF zebrafish won't occur until investigators start demanding them. And it's his sense that as experimental endpoints become more sophisticated and more subtly influenced by internal and external factors, we may be approaching a political tipping point.[6]

Consider one example: *Pseudoloma neurophilia* is a protozoon estimated to be present in half of laboratory zebrafish colonies. If too many parasites infect the brain and spinal cord and later spread to the musculature, the result typically presents as emaciation and severe spinal deformities. Clinically silent infections have been recently linked to aberrant behavior in zebrafish, representing a potentially major and unwanted complication for any behavioral studies. Other microbes besides *P. neurophilia* are being detected in zebrafish as subclinical infections.[9] Who knows what impacts they may have on delicate experiments easily perturbed by hidden pathogens? Perhaps this is the beginning of the tipping point that Chris foresaw. That's the good news.

The discouraging news is that even if there is an outcry tomorrow from scientists around the world for SPF zebrafish, the current means of preventing infections are unreliable in any large-scale zebrafish housing system. Only 10% of the water supplying tanks of fish is usually "fresh"; the other 90% is recycled through biofilters to remove nitrogenous waste and then exposed to ultraviolet (UV) light for disinfection before returning to any and all tanks. It appears there's little faith in the ability of UV treatment of tank water to eliminate unwanted microbes entirely, even though precise UV exposure intensity and duration parameters have been ascertained that can decontaminate recirculating water under controlled conditions.[10] If UV light was a truly effective and universal sterilant of recycled tank water, one presumes maintaining zebrafish as SPF would have been achieved by now.

An alternative strategy to avoid cross-contamination involves continuous flow or so-called "one-pass" housing systems, in which the tank water is still conditioned before being supplied to tanks but never recirculated. Since aqueous microbes can't swim upstream, they would be flushed out and not introduced to other tanks sharing the same water supply. But that's a very expensive and wasteful approach compared with the established practice of recycling 90% of tank water in a given housing system, and represents an added cost for which most institutions wouldn't pay. Another approach is to maintain smaller, separate housing systems so that if one isolated cluster of tanks is infected, no fish in other clusters are at risk. Similar to the continuous flow strategy, this would require more capital investment as well as more labor and reagents to maintain each of the smaller, independent systems. And there's still a risk of cross-contamination if personnel aren't careful and don't disinfect capture nets and other supplies that may be shared between systems.

What is one to do if an unwanted microbe appears and needs to be eliminated? Because there's no equivalent to the micro-isolation rodent cage for individual zebrafish tanks yet, only two options appear to have been proven effective. The first involves screening embryos by polymerase chain reaction for the pathogen(s) of concern and stocking new tank systems only with embryos that test negative.[11] The other choice is large-scale depopulation of an entire housing system, followed by chemical sterilization of all surfaces, replacing and reseeding the biofilter, and restocking the system with new fish.[12] This is not practical in most institutions and represents a serious disruption to research that most users would understandably resist.

So how do we move forward to enable science while protecting fish from complicating infections?

Two engineering advances are needed to afford the same individual fish or tank protection from infection that characterizes today's laboratory rodent housing. First, we need a process that traps water-borne pathogens in some sort of mesh filter for rapid and easy identification; one filter in-line for water from each tank, row of tanks, rack of rows, etc.—take your pick. This will provide more accuracy in knowing what pathogen(s) may be potentially infecting any fish from shared/recycled water. Over the years, I've attended scientific and trade conferences on microbe detection using biosensors, often funded by the US Department of Defense or Homeland Security out of concern for bioterrorism threats to military or municipal water sources and other population health vulnerabilities. After 9/11 and the anonymous anthrax scares of the early 1990s, money began flowing into the biosensor field to develop highly sensitive and specific means to alert authorities when water sources are intentionally contaminated. Biosensors have always been a logical technology platform conceptually because of the ability of various biomolecules to discriminate between surface proteins of highly dangerous prokaryotes. If an electrical or light signal can be amplified sufficiently whenever a receptor protein binds to its ligand or an antibody binds to an antigen, voila— you've got a nifty detection package that leverages nature's molecular specificity.

I was following this field in search of biosensors that could function in flowing air rather than flowing water. If an air-sampling biosensor was ever available, it could theoretically be adapted to monitor exhaust air from ventilated rodent cage racks and raise an alert whenever a virus or bacteria of concern was captured by the corresponding receptor molecule that was linked to some kind of electronic circuitry. But most biomolecules have evolved and continue to function in aquatic, not atmospheric, environments as a legacy to their primordial origins, so my idea for IVC's remains unfulfilled. On the other hand, large housing systems for laboratory fish seem like a much simpler and more promising starting point. Recent advances in nanowires and other atomic-scale detection devices that involve sampling aqueous fluids are exciting in this regard and may be getting us closer to rapid and highly specific fish pathogen detection.[13]

But the problem has always been what to do with a positive result? The second basic engineering advance needed is a means to remove or truly destroy circulating pathogens so that they can't spread from fish to fish. Merely re-configuring in-line UV treatment of recirculating water by increasing light intensity or exposure time sounds too obvious and probably not safe or practical enough to not be in vogue today. Instead of seeking more UV firepower, what if the same nano-filter traps mentioned above could be enhanced to capture *all* microbes flowing by? Nanofiltration has been around for almost 10 years for non-chemical sterilization of drinking water.[14] What's especially attractive conceptually is that these systems are small scale compared with their intended use in municipal water treatment plants; it's precisely that small scale that may facilitate assessment and eventual incorporation into zebrafish recirculating water systems sooner rather than later.

Finally, I mentioned earlier that zebrafish are less expensive than mice to maintain in large-scale housing systems. But by how much? My calculations several years ago indicated 20–40 times cheaper on a headcount or cage/tank count basis. These are impressive savings, alluring in an era of tighter funding for biomedical research but relevant only if zebrafish provide a suitable mouse replacement for whatever questions a scientist may be asking. The recent explosion of interest in our bodies' microbiome, enabled by more powerful genomic and bioinformatics toolkits, has led to a resurgent popularity of gnotobiotic animal models. The most popular species used to date in gnotobiology has been the mouse. This is no surprise given its dominance elsewhere and the relative ease with which it can be genetically re-engineered, thereby providing researchers with multiple options to tweak biological relationships between macroorganisms and microorganisms.

Zebrafish are no strangers to this equation, with successes described in making them axenic (albeit not for very long without at least monoculture rescue)[15] and gnotobiotic for selected portions of their resident microbes. However, one intriguing line of investigation may further increase the popularity of zebrafish in this particular application, i.e., successfully populating their pristine gut with selected microflora of *mammalian* origin.[16] If experiments such as these show that mouse-originating microbes perform the same metabolic and immunologic roles in fish as they do in mice, then why not embrace a less expensive animal model? And if part of the mouse microbiome can be recreated in the zebrafish, why not perform the same engraftment with *human* microbiome components for similarly cost-saving animal modeling of human diseases? This will be a fascinating dimension to follow if it pans out but, like much else involving laboratory aquatic animals, is dependent on solving the SPF conundrum.

# Chapter 5

# Democratize the Guide

Updates to the Guide are necessary as its subject matter changes. New knowledge arises on almost a continuous basis about the biology, behavior, care, and various uses of laboratory animals and should be incorporated in the Guide to maintain its value. This can involve the growing popularity of particular species, such as a significantly expanded section dedicated entirely to the care of aquatic animals in the eighth edition, reflecting the rise of large colonies of zebrafish. Beyond animals themselves, the Guide has adapted to changes in equipment (cubicles and later IVCs for animal housing), occupational safety practices (capturing waste anesthetic gas, prophylactic anti-viral therapy options after exposure to macaque tissues and biological fluids), supplementary federal and institutional oversight bodies (IBC's and ESCRO's) when laboratory animals are involved, and societal values expressed as additional ethical reviews (IACUC's) as these changes became mainstream.

Even though what we know and may wish to apply to the care and use of laboratory animals is frequently supplemented with fresh material, the Guide has been taking longer and longer to keep pace. The first edition of the Guide, published in 1963, was followed only 2 years later by the second edition in 1965. However, each interval between successive editions after that has been longer than the one preceding it: there were 3 years before the third edition in 1968, 4 years before the fourth edition in 1972, 6 years before the fifth edition in 1978, 7 years before the sixth edition in 1985, 11 years before the seventh edition in 1996, and 15 years before the eighth edition (copyrighted in 2011 but released December 2010).

To inaugurate the revision process, the US National Academy of Science directs the National Research Council, its subordinate administrative body, to assign this responsibility to ILAR, its subordinate administrative body for laboratory animal issues. This starts with recruiting a committee of around a dozen persons acceptably credentialed in laboratory husbandry, medicine, and use in research or testing to consider updating content where appropriate. Their proposed changes are then reviewed by a separate, similarly sized group of similarly credentialed experts to comment on these changes but, unlike the committee, don't share their observations with each other. Any major disagreements between the two groups are addressed and usually resolved before the new edition is approved for release. All of this occurs behind closed doors and is financed by organizations directly impacted by

Notes in the Category of C. https://doi.org/10.1016/B978-0-12-805070-5.00005-9

the Guide's contents, such as academic and for-profit research institutions, biomedical research societies, and various branches of the US government engaged in animal research and testing. Each revision is undertaken in an attempt to keep the Guide current and retain its relevance. So at least the more recent revisions are preceded by lots of conversations at national meetings and other venues, with folks asking each other "isn't it time for the Guide to be updated?", followed by ILAR establishing an expert panel to address this question officially. If its conclusion is "yes," then funds are solicited and the formal process begins.

Revising the sixth edition of the Guide began in 1991, with an ad hoc committee appointed by ILAR making that recommendation. A review committee of 15 comprising "research scientists, veterinarians, and non-scientists representing bioethics and the public's interest in animal welfare" was appointed in 1993 by ILAR, and the final product was issued 3 years later.[1] The subsequent revision of the seventh edition was preceded by a process for selecting committee members that, according to the committee itself, was "intensive and unprecedented for any previous version of the *Guide*. The qualifications of the thirteen members were intensely scrutinized over several months, followed by a 20-day public comment period before the committee roster was finalized."[2] Why has it taken increasingly longer to update the Guide?

A major reason is because the US Public Health Service (PHS) adopted the sixth edition of the Guide in 1985 as its primary regulatory reference document, compelling institutions that receive applicable US government research funds to follow the Guide to establish and maintain their laboratory animal care and use programs.[3] Since its inception, the Guide had always been intended to serve as just a guide (hence, its name) for good standards and practices for laboratory animal care and use. But when the US government announced it was going to require institutions to comply with the Guide and have OLAW enforce compliance via oversight and impose penalties for non-compliance (the details of which would be available to the public), the Guide instantly became a list of official do's and don'ts, at least in the United States. From that point on, every statement was to be analyzed and discussed ad infinitum (ad nauseam) for interpretation and implementation by all affected parties.

Over the years since, various guidelines in the Guide, especially engineering standards, became absolute and non-negotiable. They've dominated the conversation about what's allowed or not, mostly because they're easy to evaluate. For example, a cage holding mice weighing more than 25 g apiece either provides at least 15 square inches of floor space/animal or it doesn't. And if it doesn't, it's considered unacceptable. That's despite the origin of many of these engineering standards being a bit murky, representing a compromise between practicality and optimal animal welfare when they surfaced decades ago, and despite minimum space requirements such as those for mouse cages being published as "recommended" (Guide, page 57). Meanwhile, performance standard alternatives have remained *rarae aves* even though there's ample language in the seventh and eighth editions encouraging their consideration. It's less risky for institutions to conform to the same engineering standards everyone else uses and because OLAW serves as judge and jury to rule on the acceptability of anything out of the norm for institutions having to abide by PHS rules.

Under this reality, any further revision to the Guide is understandably a little frightening because it could conceivably change today's engineering standards. The fear is that any changes could be more restrictive or otherwise unreasonable, especially if they aren't supported by compelling objective evidence of their purported superiority. And new capital investment could be required to replace the equipment that may not comply with the new standards. One would think the logical alternative would be to keep the old stuff and just confirm that it still performs (key word) adequately, with no greater risk to animal welfare, occupational safety, or research data integrity. But the grip of engineering standards is so strong that a performance standard option remains unpopular and unnecessarily difficult to gain acceptance.

Consequently, uncertainty about changes to current engineering standards during recent updates to the Guide generates the greatest angst and battle lines get drawn. Who is appointed to the Guide revision committee, what input they solicit from whom and how much weight they give it, what recommendations they may issue, and how those recommendations would then be enforced is of considerable concern to the entire laboratory animal community. And the more years there are between Guide revisions, the longer any new engineering standards would have to be followed, even if they're later found to be wanting, until a newer edition is issued.[4] That's the situation with which we live today. It's no wonder that revising the Guide is so arduous even though its usefulness is at risk the longer it doesn't change to accommodate newer knowledge.

It's too bad that evidence-based performance standards remain so underrepresented. They have the potential not only to offer insights and breakthroughs on how laboratory animal care may be improved but also to yield significant savings in an era of tightening research funding. To the latter point, the NIH budget for biomedical research declined by over 19% in constant dollars from FY2003 to FY2015[5] with private non-profit foundations filling some but not all of the gaps. Similarly, the pharmaceutical industry's aggregate spending on research and development has essentially flattened over recent years, whether in current or constant dollars.[6]

Under these circumstances, performance standards that reduce costs and have been published in peer-reviewed journals should be more widespread. A good example involves a paper that (re)confirmed the reliability of micro-isolation caging and proper technique alone to prevent rodent pathogen transmission between cages, without wearing the full personal protective equipment (PPE) getup commonly required in barrier rodent facilities.[7] Even though the authors estimated their institution could save $150,000 per year if it switched to reduced (yet still effective) PPE, a switch from conventional PPE wasn't yet made at the time of the paper's publication. Such reluctance is widespread and particularly noteworthy because protection from inter-cage infection afforded by static or ventilated micro-isolation rodent caging, in tandem with laminar air flow hoods/changing stations, has been repeatedly demonstrated by rigorous studies for almost 30 years, with little or no evidence that extensive PPE is necessary for protection.[8] The other primary justification for continued use of extensive PPE beyond mere disposable gloves and sleeves or gowns stems from legitimate concerns about worker exposure to rodent allergens. Yet rodent allergen exposure has been known for many years to be avoidable with appropriate equipment and proper handling technique.[9] It's symptomatic of the field's resistance to change that PPE standards rightly established in the 1970s, when infectious disease was rampant in rodent colonies and before enclosed rodent cages became popular, remain largely unaltered.

Other examples abound of rigid and costly loyalty to long-standing engineering standards originating from the Guide. Take rodent cage wash and sterilization for instance. Many programs are still heating cage wash water to 180°F/82.2°C to ensure disinfection of cage surfaces even when those washed (rodent) cages are subsequently sterilized in an autoclave or dry heat oven prior to reuse. Why? Because that's a temperature that's still published in the Guide even though it's surrounded by explanations and options (page 71).[10] By contrast, my former program re-configured tunnel washers to use only room temperature water and no detergents, resulting in rodent cages coming out plenty clean prior to sterilization and a much more comfortable cage wash area in which to work.[11] Another large and established program autoclaves only rodent bedding prior to filling cages; dirty cage parts and water bottles are sanitized without subsequent sterilization and then filled with autoclaved bedding. That approach has been sufficient to protect the negative infection status of a mouse colony with an average daily census of 40,000 cages. Think of the labor hours, water and energy consumption, and safety risks avoided with no cage sterilization, not to mention the capital investment and floor space spared by not installing big box sterilizers. And if there was ever an outbreak involving an excluded pathogen, chemical sterilants would work just as well on a temporary basis to eradicate the unwanted bug.

Another holdover from an earlier time: in animal housing rooms, it's still common to wash animal room walls and ceilings with disinfectant on a regular schedule even when bioluminescent detectors or agar culture plates show insignificant levels of biological residue on these surfaces.[12] Even in the absence of such measures, if the walls and ceilings merely look clean, don't come in contact with animals or their waste, and the animals are reliably free of infectious disease, why wash those walls and ceilings until there's a good reason to do so?

Yet one more and still dominant line of thinking: because the Guide states that the acceptable range of relative humidity is considered to be 30%–70% for most mammalian species (page 44), people in the know still apply that range to rodent rooms filled with micro-isolation cages, as if those cages were still fully exposed to room air. Thus, many facilities in northern US climes continue to invest in and maintain expensive humidification systems for those rooms because outdoor relative humidity levels can drop below 10% in the winter. But we've known for some time that micro-isolation cages provide an internal humidity "cushion" of up to 10%–15% of additional humidity inside the cage, thereby providing its inhabitants with an environment that's still compliant with the Guide even when the room is "too dry".[13] A program with which I'm familiar avoided spending tens to hundreds of thousands of dollars to upgrade its HVAC infrastructure for those few frigid days a year when the animal room humidity levels were in the twenties, precisely because the rodent cages' internal environment remained within the humidity range stated in the Guide.

An even bigger opportunity than any of the above for cost savings involves the conventional practice of changing rodent cages at least every week or two. The current Guide retains language from past editions on this subject, such that "In general, enclosures and accessories, such as tops, should be sanitized at least once every 2 weeks. Solid-bottom caging, bottles, and sipper tubes usually require sanitation at least once a week." (page 70).[14] This engineering standard has become so entrenched that every rodent cage on a rack or in an entire room is often faithfully changed every 1–2 weeks regardless of the degree of soiling inside each of those cages. It's considered safer from a compliance point of view as well as a simpler vivarium management tactic to replace occupied cages with clean ones according to a calendar rather than according to the need. Even if a new shipment of mice or rats arrive on a Monday, if their room is scheduled for a change on that Wednesday then their cages get changed that Wednesday. This avoids any uncertainty about when the animals were last placed in a clean cage and everyone's assured that the calendar standard isn't violated.

But many cages contain fewer than the maximum number of mice or rats allowed, and their bedding can go longer than 1–2 weeks before it needs changing, especially if IVCs are involved and the animals do just fine.[15] Some programs have demonstrated that for any given 1- or 2-week period, one-third of mouse IVCs scheduled for a change aren't soiled enough yet to justify a change, based on the correlation between the visual degree of soiled bedding in a cage and its ammonia level.[16] How big a deal is that? When I was at MGH we explored a performance standard option for changing cages when we had an average daily census of 27,000 mouse cages with 85% being ventilated and 15% being static. It was calculated that if cages were changed only if and when needed rather than at least every week (for static) or 2 weeks (for IVC), we could delay or avoid more than 240,000 cage-changes a year while freeing up over 8000 labor hours (the equivalent of four full-time employees) to be re-assigned to less physical or more critical tasks. For my current program where we wash and sterilize 7000 individually ventilated mouse cage bottoms a week, a 30% reduction in throughput could save up to 14 tons of bedding and

14,000 gallons of water a year. Of course, one needs to account for all pertinent inputs, such as the type and amount of bedding, size and shape of cages, air exchange rates, and animal density per cage, in considering when cages need changing. Nevertheless, fewer changes or longer intervals between changes also serve to disturb animals less often, which is desirable for them as long as the micro-environment isn't too dirty. Plus the timing of cage changing or reducing its frequency may avoid discrepancies in research data.[17]

Would the Guide permit such an approach? The same section that contains the 1- or 2-week calendar standard also provides the following statements: "The frequency and intensity of cleaning and disinfection should depend on what is necessary to provide a healthy environment for an animal" (page 69); "Soiled bedding should be removed and replaced with fresh materials as often as necessary to keep the animals clean and dry and to keep pollutants, such as ammonia, at a concentration below levels irritating to mucous membranes" (page 70); and "There is no absolute minimal frequency of bedding changes; the choice is a matter of professional judgment and consultation between the investigator and animal care personnel" (page 70). Sounds like an okay to me to exercise one's judgment rather than to follow a calendar blindly.

How would such a performance standard work in reality? Does the Guide provide any advice? On page 71, it states that "…decreased sanitation frequency may be justified if the microenvironment in the cages, under the conditions of use (e.g., cage type and manufacturer, bedding, species, strain, age, sex, density, and experimental considerations), is not compromised" and "Verification of microenvironmental conditions may include measurement of pollutants such as ammonia and $CO_2$, microbiologic load, observation of the animals' behavior and appearance, and the condition of bedding and cage surfaces." Thus, there is an assortment of metrics offered by the Guide, in combination with professional judgment, to apply a performance standard for changing cages. Conveniently, animal care staff evaluate "the condition of bedding and cage surfaces" every day to identify cages that need changing *before* their scheduled 1- or 2-week change (e.g., excessive urine in bedding from diabetic inhabitants). These are known as "spot changes" (so-called because cages are changed on-the-spot or perhaps because they're changed as soon as they're spotted). If technicians are already entrusted to identify excessively soiled cages and then change them immediately upon detection, the only difference between the engineering (calendar) standard and the performance (change when needed) standard is certainty versus uncertainty. In other words, I know that every cage in a given room is scheduled for a change on a specified day (engineering) versus I don't know how many cages in a given room will need changing today, later, or never but I will change any and all cages that need changing as soon as they're discovered (performance). Under the performance standard, every day would be (only) a spot change day for the entire facility. However, the resultant loss of chronological predictability makes vivarium managers very uncomfortable because they can't organize the workday and dirty (or not) cage throughput as neatly as before. Nonetheless, I'd swap that predictability any day in a minute for an overall one-third reduction in cage wash throughput and its resultant benefits. Stated differently, anyone in a senior oversight role for program expenditures who doesn't at least consider abandoning the calendar standard is violating his or her fiduciary responsibilities.

Some common inefficiencies don't involve engineering standards at all, but arise from unnecessary internal compliance processes. For example, many IACUC's conduct annual reviews for all active protocols. But only a small portion of these protocols may actually require annual reviews, as determined by the species (those covered by the AWA) or funding agency (e.g., US Department of Defense or DoD) involved. By contrast, NIH-funded protocols require review only every 3 years. In some places, the annual "review" for non-AWA, non-DoD protocols involves a simple checklist or IACUC administrative staff attention. But the investigator must still fill out and submit something when it's unnecessary. On a related note, a few years ago OLAW allowed some "significant" changes, with USDA concurrence, to protocols to be reviewed administratively and, without full IACUC involvement, provided that "a veterinarian authorized by the IACUC" is consulted.[18] Yet from what I've heard at compliance conferences and elsewhere, many IACUCs haven't adopted this shortcut.[19] Self-imposed regulatory burdens such as these consume too much valuable time and could be eliminated.

Also beyond Guide engineering standards are new alternatives to traditional means of monitoring rodent colonies for unwanted infections. The usual strategy has been to transfer small amounts of soiled bedding from cages of research rodents to cages of uninfected "sentinel" rodents, with the expectation that any microbes of interest such as unwanted pathogens would infect the sentinel rodents that could be tested later. In earlier times, one would look for illness or an antibody response to the microbes of concern in these sentinels. These endpoints were later augmented or replaced with ultrasensitive tests for microbe DNA. But we've known that transmission of a given virus, bacteria, or parasite from one rodent to another, even when they're housed in the same cage, is not always reliable. More recently, it's been shown that swabbing inanimate surfaces for microbial DNA is just as or more accurate as swabbing sentinel mice. These inanimate surfaces can be air exhaust plenums on IVC racks, bedding dump stations, entry air shower dust traps, etc. Similarly, there's good evidence that swabbing laboratory mice arriving to a vivarium from non-commercial sources (known in the vernacular as mouse "imports") is just as, if not more, reliable than employing sentinels and soiled bedding to determine whether the incoming mice are carrying excluded pathogens.[20] My program abandoned sentinel rodents for routine colony health surveillance a couple of years ago and switched to swabbing air exhaust plenums and imported mice directly. This spares 1600 sentinel mice and rats a year at an annual savings of $11,000 in animal procurement costs. In addition, investigators requesting mice imported from academic collaborators can get access to them within 2 weeks after arrival if tests are negative, rather than 6–9 weeks of waiting to see if the sentinels got infected and for precautionary anti-parasite treatment to take effect, as used to be the case. These and other alternatives to engineering standards and compliance processes can improve a program's financial bottom line. But what impact could they make collectively in the grand scheme of US biomedical research funding? Let's look at the numbers. In FY16, NIH awarded over $17 billion for both competing and non-competing Research Project Grants (RPGs), including 6010 new or renewal (competing) R01-equivalent awards at an average of $458,287.[21] From public and private financial data to which I've been privy over the years, an average of 1.7% of an academic institution's research budget is spent on direct (operating) expenses for laboratory animal care alone,[22] equating to $291 million within these FY16 NIH awards. It's been my experience and observation that, typically, at least 20% (actually closer to one-third, but let's be considerate) of a program's activities and expenses are wasted on costly engineering standards or conventional ways of performing husbandry, veterinary care, administration, staff training, etc. If these costs were eliminated on a national scale, more than $58 million in savings could be redirected to 127 additional new or renewal (competing) R01-equivalent awards of the same average size.

Or perhaps there are even greater needs to address in the realm of biomedical research, such as the graying of US academic science. To wit, the proportion of R01-equivalent principal investigators aged 35 years and younger dropped from 21% to under 2% between FY80 and FY14, whereas those aged 66 years and older increased from less than 1% to over 9% over the same interval; an almost identical trend was seen for the RPG Director awards.[23] Scientists, like others, are living longer and working longer, which by itself is fine. But research funding has not kept pace with the growing cost of science or the growing population of junior researchers. It's logical that, everything else being equal, well-established scientists have a competitive edge for research grants because they've had more time and opportunity to build on earlier progress, both personal and communal. And if research grant applications are reviewed solely on their merits, as they should be, then more experienced scientists are likely to be at an advantage. But we jeopardize the future success of US science if younger researchers are unable to establish independent research careers after the baby boomers eventually leave their laboratories. NIH recognized this vulnerability by establishing blocks of grants for those who have not yet won an R01 award, such as the Director's New Innovator Awards (DP2s), Early Independence Awards (DP5s), and Career Transition Awards (K99's). Recent data about this concern, while recognizing that younger ages may or may not be involved, are encouraging; between FY11 and FY15, 35% of all R01 awards were made to new investigators.[24] On a related front and at the time of this writing, NIH is considering how to cap research awards so that the biggest laboratories (receiving the biggest grants) don't squeeze out smaller players such as new and mid-career scientists.[25]

What does this have to do with performance standards of laboratory animal care? NIH awarded 545 DP2, DP5, and K99 grants in FY16 totaling over $211 million, at an average $387,369 per award.[26] If laboratory animal care programs in the United States replaced their costly engineering standards with more efficient performance standards and eliminated other unnecessary expenses, the $58 million in savings projected above could be rerouted to 150 additional new investigator awards such as these. Who knows what new discoveries and medical advances could be realized if this is achieved?

Why aren't performance standards and operational efficiencies more popular? It's not due to a lack of knowledge; there are lots of very smart and highly educated persons who are avid readers of laboratory animal journals and trade publications and can understand the rationale for various performance standard options. I think there are four other explanations for why engineering standards remain unshakable, and these are related to more universal elements of resistance to change, irrespective of how strongly change may be needed or how attractive the changed outcomes may be. The first is the lack of a perceived need for change. Until recently, research funding was abundant and there was no financial pressure to reduce expenses. Those days are over and institutional insistence on cutting costs will only increase in the years to come. The second is the lack of imagination. The practice and management of laboratory animal care is highly imitative as well as conservative, and until now, there was little incentive to innovate outside of established norms. We haven't had to be very resourceful or creative when it came to vivarium operations and veterinary care, especially on a colony-wide basis; so these talents have atrophied or have been forgotten for these applications. The third is the lack of training or experience in managing change itself. Many business school courses and management books are devoted to this subject because most other industries experience periodic if not constant threats to their status quo. Only by thinking differently, trying out new things or actions, and successfully implementing changes can organizations in these industries survive, if not thrive. But strategies and tactics for managing organizational change might as well be in an obscure foreign language when it comes to laboratory animal care programs; they've never been in our vocabulary.

The fourth explanation for resistance to performance standards is emotional, involving perhaps the most primitive and powerful emotion, namely fear. At a personal level, this is the fear of getting blamed, disciplined, or fired if a "new" way backfires and causes greater problems. The first three factors listed above will be overcome only when the fear of losing one's job because costs aren't reduced exceeds the fear of losing one's job because of other problems that arise while pursuing cost reductions. Until that time arrives, the fear of changing anything will continue to dominate. At a communal level, fear of a different sort, i.e., peer disapproval by not conforming to what are erroneously proclaimed to be "best practices" (an overused phrase often misunderstood or misinterpreted to mean "the only option allowed"), is in play. So while an individual's fear of getting punished for mistakes may represent the ultimate refuge from change, there may be an even stronger and subtler driver for the reluctance to be different from others in one's tribe.

Decades ago, Paul DiMaggio and Walter Powell of Yale University coined the perfect term for why groups trend toward conformity, i.e., institutional isomorphism. The first wave of organizational homogenization was competitive isomorphism, in which capitalism's ruthless pursuit of efficiencies in the industrial revolution and early 20th century drove organizations to a shared structure and bureaucracy that came to eclipse alternative models. The rise of the state and of professions in the latter half of the 20th century similarly drove organizations not engaged in for-profit enterprise to converge to eventually domineering models of bureaucracy and thinking. Quoting from the authors, "Once disparate organizations in the same line of business are structured into an actual field (as we shall argue, by competition, the state, or the professions), powerful forces emerge that lead them to become more similar to one another. Organizations may change their goals or develop new practices, and new organizations enter the field. But, in the long run, organizational actors making rational decisions construct around themselves an environment that constrains their ability to change further in later years."[27]

In their essay, DiMaggio and Powell identified three mechanisms by which institutional isomorphism occurs. Coercive institutional isomorphism is characterized by "formal and in-formal pressures exerted on organizations by other organizations upon which they are dependent and by cultural expectations in the society within which organizations function." These pressures can originate from laws, regulations, and their standardized interpretation as well as more self-imposed standards of conduct and outputs. The second means by which institutional isomorphism occurs is normative, which is mostly a consequence of professionalization. In this version, the authors posit that (advanced) formal education legitimizes a particular knowledge base which is then strengthened by professional networks of experts who possess such education and knowledge. These networks in turn "create a pool of almost interchangeable individuals who occupy similar positions across a range of organizations and possess a similarity of orientation and disposition that may override variations in tradition and control that might otherwise shape organizational behavior." The more experts produced and the more they interact, the less likely those experts are to imagine or tolerate ways of doing things that defy the professional consensus. Mimetic institutional isomorphism is the third process, usually found during times of instability in a given field. When everything outside the organization is changing at a rapid pace, it's cheaper and less complicated to imitate how another organization is assembled and managed rather than try to invent or revise an organizational arrangement yourself. This allows you to focus on battling dynamic externalities without internal objections because you're organized similar to other combatants.

It's not a stretch to see how coercive and normative mechanisms of institutional isomorphism may be operative in laboratory animal care, applicable to both individual research institutions and professional organizations that promulgate or rely on the AWA and the Guide for acceptable standards and practices.[28] In fact, DiMaggio and Powell's sociological construct and the subsequent works of many others in their field may offer a novel prism through which to understand continuing allegiance to engineering standards and widespread resistance to performance standards. Even better, scholarship about institutional isomorphism in laboratory animal care may provide insight on how to overcome that resistance, either within one's program or industry wide.

In the meantime, what's to be done to leverage the promise of performance standards to the benefit of many? As hinted in the Guide and accepted as gospel elsewhere, performance standards must be evidence-based in order to pass muster with both the local IACUC and the community at large. Ideally, that evidence should be generated from a sound scientific design, quantified, and subjected to statistical analysis whenever possible for greater credibility (NB: contrary to the opinions of some, publishing one's performance standards outcomes as a peer-reviewed paper in a reputable journal is not required for these outcomes to be accepted by the local IACUC involved or be acceptable to other parties). The second most important factor in advancing performance standards is a means to share and challenge ideas and experiences about various performance standards. Otherwise, awareness of what others have attempted and adopted requires one to attend a conference or visit another facility personally in order to learn about them. Having a robust information exchange is especially important since there's perceived safety in numbers; the more performance standards that are out there for analysis and discussion, the more comfortable everyone feels. Conversely, if only a few performance standard options are in circulation and aren't appealing to enough programs, then it's too easy for everyone to step away from the general concept and retreat to the comfort of the status quo and engineering standards.

No such organized social venue or platform dedicated to performance standards currently exists. To fill that void, I have proposed establishing a web-based repository of them for everyone in the field to access.[29] Such a virtual repository of performance standards would have the following features:

- a reliable, secure host and server to withstand hacks and to safeguard content
- a user-friendly presentation with easily searchable subject matter
- entries tabulated using a multitude of categories, perhaps mirroring the Guide's Table of Contents
- entries linked to pertinent sections of the Guide, AWA, and other regulatory documents where applicable
- password-restricted access, at least initially (more on this later).

Performance standards could be described and submitted in any format that one chooses (imposing an inflexible standard submission form risks stifling good ideas). Only two requirements would be enforced: all submissions must have already been approved by the submitter's local IACUC or equivalent, and the supporting evidence for the performance standard must be credible, ideally generated using sound scientific principles with appropriate data analysis if applicable. Entries would undergo two filters prior to posting in the repository. Each filter would be performed by knowledgeable volunteers or paid employees, perhaps eventually replaced by intelligent software. The first filter would detect and reject gibberish or venomous diatribes. The second filter would be more sophisticated but still rudimentary with respect to acceptable criteria, again to avoid discouraging submissions because they weren't sophisticated enough. Instead, the second filter, as a review panel composed of peers, would identify obvious content gaps or errors that the author would have to address satisfactorily before the submission could be posted on the website. Beyond that, everything would be acceptable. Authors would need to understand that once posted, their submissions could be challenged or dissected, with readers able to post questions and commentary (all of which would be subjected to the same two filters) and linked to the original submission in the database and on the screen. Readers would need to understand that submissions and subsequent annotations may not be applicable to their respective programs. In a similar vein, just because a performance standard is posted doesn't imply its acceptability to other programs' IACUCs, accreditors, and regulators.[30] Each posting would have to be analyzed through the lens of each program and institution.

Another potential benefit from this repository is facilitating collaborative investigations between multiple programs. Say a performance standard that's posted is intriguing to another program that feels there are a couple of gaps that need to be addressed. These gaps could involve how the posted performance standard was designed or implemented, how applicable data were generated or analyzed, or other minor concerns. Or perhaps another program wanted to pilot that performance standard in its facility but to alter the details, such as type of caging or bedding, inhalant rather than injectable anesthetic, or guinea pigs instead of hamsters. But rather than start from scratch, either of the second parties could leverage the knowledge gained by the author of the original posting and get a faster start in extending or refashioning that performance standard to their particular liking. Or if the author possesses an expensive machine or a complicated laboratory assay that could be shared to investigate variations of that performance standard with other programs, more parties could profit at less cost and in less time. The beauty of a widely available repository is in capturing these conversations as well as the results of the collaboration for the benefit of all. If a published paper from such collaborators was the outcome, the repository could simply link it to the original submission along with the ensuing discussion, just like any topic stream in other social media. The result would be a "bottomless" trove of sample performance standards, dialogues, revisions and enhancements, and testimonials, and more that would be archived, easy to mine, and grow as we learn more about laboratory animal biology, behavior, care, medicine, and models.

Perhaps a good example of how this could all work is provided by the naked mole rat, *Heterocephalus glaber*. This species has grown in popularity as a laboratory animal due to its unique energy metabolism, resistance to pain and cancer, longevity, and social behavior compared with other mammals. It was even selected in 2013 by *Science* magazine as the "Vertebrate of the Year."[31] The natural habitat for this animal is underground. Even though it can be maintained in vivaria under more conventional conditions,[32] would a natural subterranean environment be more conducive to it expressing natural traits that are so scientifically attractive? If so, how could such a habitat be maintained and still comply with the Guide? A multitude of performance standards could provide a multitude of options, each of which may be a better fit than others for a given institution's researchers and program of laboratory animal care. This is a much better approach than the current practice of anointing one peer-reviewed publication as a "best practice," to the likely discouragement of other options that may be better in light of differing circumstances yet entirely acceptable. One is reminded about papers of yore describing how nine-banded armadillos (*Dasypus novemcinctus*) were maintained as research animals for studying leprosy and monozygotic quadruplets.[33] How would the housing arrangements (using unsealed plywood!) have fared under today's insistence on pristine environments? Looking forward, what about other novel animal models that fall outside established norms? How should we react to a recent paper that described how mice captured from barns or purchased from pet stores, with their many attendant murine pathogens, replicate the adult human system better than the SPF laboratory mice for which we've worked so hard for decades to establish.[34] If these findings are confirmed and if "dirty" mice like these are deemed worthy of further investigation, I'm afraid many programs will forbid their presence to avoid infecting the rest of the colony rather than consider alternatives. But our charge is to enable, not impede, biomedical research. So we're obligated to find ways to isolate such animals from the rest of the vivarium without breaking the bank. A robust dialogue in a performance standards repository could offer a variety of housing and husbandry possibilities.

From where would performance standard postings and commentary come? Initially, it's appropriate to limit the repository's reach to those who are engaged in laboratory animal care and use. Familiar and trusted industry-wide organizations dedicated to information sharing, such as ILAR, AALAS, PRIM&R, or SCAW, could provide a logical base for the repository while safeguarding its objectivity and neutrality. But eventually it would be wise to open the repository up to literally everyone. It's well established that better approaches sometimes surface from non-experts and unknowns. Crowdsourcing for new solutions to particular problems or needs is becoming commonplace, whether soliciting input from individuals or via focused group exercises (otherwise known as hackathons).[35] Why would we think we always know the best or would have a monopoly on alternatives to do things better? Opening up the repository to the entire planet could attract engineers, materials scientists, mathematicians, product designers, statisticians, and information technology (IT) specialists from other industries that had rarely, if ever, been linked to ours in the past, as well as wildlife conservationists, livestock production scientists, and other experts from disciplines more aligned with captive animal biology and care. Vendors in our industry could pose questions or offer rewards for solutions to vexing product issues in a dedicated corner of the repository, as is done in other industries.[36] Concern about verbal or IT attacks (or worse) by cranks and extremists is valid and not to be minimized. But there are effective defensive strategies out there for this sort of thing that could be applied to the repository where appropriate. The opportunity to accelerate progress in our field by inviting in the public dwarfs the risks of keeping the doors shut tight.

Chapter 6

# Laboratory Animals and GDP

Over the past 30 years, many formerly poor countries have seen their fortunes improve (literally) by relaxing domestic fiscal controls, adopting market-driven business policies, and increasing economic ties with other nations. The prosperity resulting from these rises in gross domestic product (GDP) and related indicators in so-called low and middle income countries (LMICs) has usually increased the disposable income of their citizens who in turn spent some of their new money to increase the amount of previously unaffordable animal protein in their diets. This relationship between average personal incomes and sources of food calories is known as Bennett's law, originally based on conclusions drawn about the relationship between wheat and other foodstuffs as nations' circumstances change, and has been validated since by other agricultural economists.[1] Increased demand for dietary animal protein has led to increased food animal production in these countries to supplement animal food imports.[2]

A second but less popularized relationship between rising GDP and animals is the increase in pet ownership and spending on pets as citizens enjoy more disposable income. One analysis estimated changes in the number of pet dogs in 53 countries between 2007 and 2012, showing that the countries with the fastest growth were all LMICs, starting with India (58.1%, ending with 10.2 million dogs) followed by the Philippines (38.3%, 11.6 million), Venezuela (29.8%, 3.1 million), Russia (28.0%, 15.0 million), Argentina (20.1%, 9.2 million), and Brazil (14.3%, 35.7 million), versus the United States (2.2%, 75.8 million) and the European Union average (−0.1%, 2.8 million).[3] Financial projections for the global pet care market predicted that over 40% of its retail value gains over a recent 5-year span (2012–17) would be driven by LMICs, led by Brazil, and then Russia, Mexico, and China.[4]

What other connections may exist between animals and changes in a nation's wealth? I suggest a third category besides food or companion animals, namely laboratory animals, driven by government investment and private capital rather than consumer spending. Not only have personal incomes risen in step with the improving GDP in LMICs over the past 30 years but also spending on science. As countries become richer, they acquire the wherewithal to establish and maintain expensive research infrastructure. To wit, the largest percentage increase in a country's gross expenditures on research and development (GERD) from 2003 to 2013 occurred in nations in Eastern Europe, Asia, and Latin America.[5] The most recent GERD data from the Innovation Policy Platform, a joint initiative

Notes in the Category of C. https://doi.org/10.1016/B978-0-12-805070-5.00006-0

of the Organisation for Economic Co-operation and Development (OECD) and the World Bank, include the expected leaders (United States, Canada, much of Western Europe, Japan, South Korea, Taiwan, and Australia), who are joined in the top category by Brazil, China, India, and Russia.[6] The major scientific journals now publish paid advertisements or special issues that highlight the intellectual strengths and ambitions of a particular research institution in a LMIC or the entire LIMC itself. The same journals also issue objective analyses that track research and development expenditures, citations of peer-reviewed publications, and issued patents in various parts of the developing world.[7]

Much of this spending is devoted to life sciences, with hopes of discovering better treatments for endemic diseases (particularly if these diseases don't get sufficient attention elsewhere), attracting private investment, and luring expatriate scientists and physicians back to their native lands with offers of generous funding, state-of-the-art laboratories, and less costly clinical research settings. While these details are often touted in the public and business press, corresponding evidence of increasing numbers of laboratory animals are not as easily found, no doubt due to political and security fears (notable exceptions are media accounts of the rise of laboratory monkey production and research in China while its scientists assume a leadership position in genetically editing these species[8]). In addition, many LMICs don't have established or credible registries for tracking the number of laboratory animals or even a directory of organizations that breed or use them domestically. Regardless, there are other indicators that more wealth and more life sciences research and development in LMICs correlate with higher numbers of laboratory animals.

One reliable metric is the increase in the number of AAALAC–accredited units outside the United States. Even though some nations have their own regulatory and voluntary oversight mechanisms, AAALAC accreditation has become an accepted standard of high-quality laboratory animal care and use for research institutions, commercial vendors, medical R&D companies, and other entities around the world. The total number of accredited units rose from 692 to 952 from 2005 to 2015, a 38% increase. The United States and Canada accounted for the largest number of units each year but at the smallest rate of increase in total units at 8% over these 11 years. By contrast, European units increased by 229% and those along the eastern Pacific Rim and in south Asia rose almost 10-fold. In fact, the eastern Pacific Rim and south Asia generated over half the increase in accredited units during this time (see chart).[9]

At the time of this writing a year and a half later, 968 accredited units were listed on the AAALAC's website.[10] Of these, 671 (69%) are in the Unites States and Canada, 181 (19%) along the Pacific Rim (including south Asia), 100 (10%) in Europe, 9 in Latin America and the Caribbean, and 7 in the Middle East and Africa (including Mauritius). More to the point of this chapter, 124 (13%) accredited units have been located in LMICs, compared with less than one-tenth of this number 15 years earlier. While some of these units belong to multinational corporations or the US military, the fact that they are sited in countries

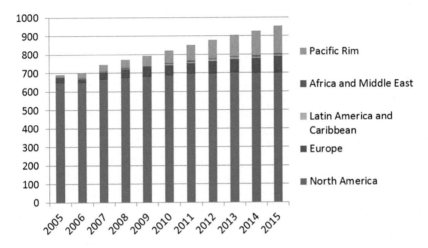

AAALAC-accredited institutional units over time, by geographical region. *(AAALAC International.)*

previously not populated with multiple, if any, accredited units is strong verification to the growth of life science endeavors in these locales. China alone
has 67 units, a remarkable figure considering that there was only one or two as
recently as 2007, and a testament to the recognized value of AAALAC accreditation in countries without long histories of established laws and regulations for
laboratory animals. Just as impressive, Shanghai is home to the most accredited
units (17) of any city in the world, followed by San Diego (15), Boston (14),
and Beijing (12), with Cambridge (Massachusetts), Houston, New York, and
Philadelphia at 11 apiece.

Tracking the growth of the laboratory animal industry in LMICs doesn't
have to be limited to publicly available information. Beyond trends in AAALAC
accreditation, one could inquire about changes in sales figures for laboratory
animals, supplies, and equipment in various countries. Selected manufacturers of rodent ventilated caging, all with a global presence, generously shared
their proprietary sales data for this chapter with the agreement that the firms
would remain unidentified. Let's consider the number of IVC racks sold for
rodent barrier housing in vivaria in China, India, and countries in the Middle
East, Latin America, and Eastern Europe. When comparing the 5-year spans
of 2002–06, 2007–11, and 2012–16, the total number of racks sold was 308,
542, and 981, respectively. Thus, each interval represented almost a doubling
in IVC rack sales over the prior interval. While admittedly a small sample
size looking only at a narrow slice of the industry, these numbers nevertheless
provide more evidence of the growing housing (and use) of laboratory rodents
in LMICs.

Another measure, although less quantifiable, is the increasing amount of
attention paid to acceptable standards of laboratory animal care and use in

LIMCs expanding their life sciences as their GDP rises. How to harmonize these standards amongst nations of varying economic status while respecting local cultures and traditions has been the subject of international workshops on laboratory animal use[11] and veterinary medicine.[12] A survey of the state of the 3Rs in public and scientific circles in Brazil, China, and India was recently published. The authors found that while there is still much to do, national agencies and professional scientific societies recognize the universality of the 3Rs and are earnestly incorporating them into their guidelines, nascent regulations, and protocols.[13] A year after this survey was published, China released proposed regulations for laboratory animals, the first for this country at the national level.[14] This reminds me of a lecture I gave over 12 years ago at an academic medical center in Ürümqi, the capital of Xinjiang Province in the far northwest corner of China. In the course of fielding questions after speaking on the use of animal models in research, I acknowledged the raised hand of a young woman at the back of the hall to hear her petite voice ask in perfect English, "What about the 3Rs?" I was both embarrassed (since I had neglected to include that slide from another lecture at the same hospital a year earlier) and elated to hear someone from the younger generation in a relatively remote region of the planet inquire about my laboratory animal ethics.

I raise the possible linkage between a country's economic state and its use of laboratory animals not as a mere factoid or curiosity. Rather, I'm intrigued about the number of laboratory animals and associated economic statistics in our industry, such as employment, capital investment, square footage or meters, etc., as indicators of the vitality of the entire life science picture for a country. Certainly, lots of other kinds of expenditures and activities are in play in a growing, static, or declining life science sector. But tracking laboratory animal-related metrics may provide early warning or later confirmatory signals of larger scale trends. And since shifts in life science investment on a national scale mirror the changes in a country's social and economic health, could metrics of laboratory animal use be informative in a more general sense? One is cautioned here by the old adage, "just because you have a hammer doesn't make every problem a nail." I don't mean to oversimplify or look only through a laboratory animal lens. But investigating the economic and social correlations of all these moving parts by means of animal research in a particular country may be enlightening.

In the meantime, the expansion of animal research in LMICs raises opportunities and challenges. As vivarium personnel in these countries become more connected with peer institutions and organizations elsewhere, their familiarity with endemic species may be invaluable as new animal models are needed for emerging diseases. Take Zika virus, for example. The common marmoset (*Callothrix jacchus*) has been shown to be naturally infected with Zika in northern Brazil although its role as a possible reservoir for the virus and its susceptibility to actual disease following infection have not been established.[15] If they are permissive hosts for infection, marmosets in particular could be an intriguing model species to study the interplay between virus, host immunity, and host genetics. Knowledge gained from these studies could be helpful for developing safe and effective vaccines, drugs, and other interventions to protect public health as well as to protect human fetuses from the devastating cases of microcephaly associated with Zika infection. Mouse models of Zika infection in utero have been established,[16] but the placenta of rodents differs from that of primates, so questions may arise that could be answered by using a primate model. The marmoset genome has been sequenced and gene editing has been successful in this species.[17] When bred, marmosets usually produce twins. Thus, one could edit genes suspected of influencing viral teratogenicity in one fetus while its twin serves as a control (unless fetal chimerism creates too much background "noise"). The traits and the research options they offer have increased the demand for laboratory marmosets in the United States to such a degree that there is a shortage of available animals, even before considering any Zika research. By contrast, marmosets are quite common in Brazil, even within the city limits of São Paulo and elsewhere, and research facilities have the know-how to breed and keep these animals healthy in their native environment. So rather than considering Brazilian vivaria merely as suppliers of laboratory marmosets to the United States, why not establish scientific collaborations on Zika and other urgent initiatives that take advantage of local expertise in husbandry, breeding, and veterinary care while leveraging the talents of molecular biologists, virologists, immunologists, and reproductive pathologists in both countries?

Transnational collaborations such as this should become more common as life science activities grow in LMICs. Animal research programs there can provide more animals at lower costs with staff scientists ably knowledgeable and skilled. However, what we in wealthy countries take for granted with respect to laboratory animal health and medicine can sometimes be lacking in less endowed settings. Six years ago I visited an established scientist in a country that shall go unnamed. His institution had just opened a new mouse imaging core that included scanners for advanced and expensive modalities, competitive in every way with mouse imaging cores I had seen at major research institutions in the United States. This scientist had been awarded a research grant from an international foundation that totaled seven figures in US dollars. But he had to return the money because he couldn't keep his mice free of common murine viruses, an essential requirement for his research project. The vivarium at this institution was physically more than adequate to protect these mice but the husbandry practices and veterinary expertise were deficient. And I've seen this in other LMIC settings—gleaming new vivaria with the latest rodent IVCs and large animal housing, surgical suites, and cage wash areas are built to showcase governments' ambitions to become life science powerhouses but there's often inadequate investment in the human capital necessary to operate these facilities properly. That's why it's important that there be an abundance of encounters and exchanges involving laboratory animal veterinarians from all nations.

In that vein and also because it would be fun, I organized a Latin America Laboratory Animal Science Fellowship program in partnership with the International Council for Laboratory Animal Science (ICLAS) when I was at MGH. Representatives of ICLAS would solicit applications from all over Latin America and we would select the fellows together. My program provided funding to allow one fellow at a time to live in Boston for 2–3 months while they were embedded in our program to learn how we approached vivarium management and veterinary support. Over the course of 2 years, we hosted six women from five countries (Argentina, Brazil, Chile, Costa Rica, and Uruguay). Five were veterinarians and one was a vivarium manager. Each fellow was required to give our leadership team two presentations and also undertake a self-selected project intended to benefit her program back home. The first presentation, given shortly after the fellow arrived, was a description of her institution's program of laboratory animal care and use while the second presentation at the end of the fellowship was a summary of what she had learned from us plus a status report on her project. The other requirement expected of our fellows was that they maintain contact after they left and share any vivarium management insights from their own programs with us.

That's because exchanges such as these should not be a one-way street. Too often, those of us from wealthy countries insist that our way is the only way, and impose expensive engineering standards when they are neither necessary nor affordable in LMICs. Quite to the contrary, having fewer resources available is often a driver for innovation and entrepreneurship in today's emerging economies.[18] I figured our Latin America fellows were more likely than us to identify cost-savings opportunities since their programs weren't funded as generously. So after we showed them it was perfectly okay not to follow US engineering standards lock-step and how to evaluate alternative performance standards, I was eager to learn about any operational improvements that they devise back home for possible use in our program.

Finally and from a different angle, can laboratory animal numbers and investments be a gauge for changes in a society's culture separate from changes in its economy? As mentioned above, rising GDP is affiliated to some degree with rising pet ownership. However, pet ownership trends in LMICs are being driven not only by growth in personal incomes but also by demographics. More persons in these countries are moving to and living in urban areas with smaller housing spaces that in turn are more compatible with smaller pets. In addition, rising personal income has correlated with dropping birth rates in LMICs. All these developments, perhaps accelerated in China by its one-child policy, mean more owners are relying on smaller animals as loved companions.[19] Thus, the advent of the phrase "pet humanization" as a growing cultural and commercial phenomenon in both wealthy and developing countries.[20]

Will these same pet owners become sympathetic to anti-animal research propaganda that's broadcast in just as shrill a tone as in developed countries? When our Latin America fellows at MGH would present an overview of laboratory animal research in their countries, one theme heard from all our guests was the rise of animal extremists in their countries, along with a disappointingly absent or muted counter-campaign to educate the public and politicians about the need for responsible use of laboratory animals in the foreseeable future. Is the global village, especially amongst younger generations, so well linked that the battle for hearts and minds for the responsible care and use of laboratory animals is already lost in LMICs that simultaneously are expanding their life science footprints? Or is there still time for (animal) research advocacy groups in developed countries to advise and assist institutions and organizations in LIMCs before protests and other anti-research activities have a worse impact?

# Part II

# Laboratory Animal Medicine

# Chapter 7

# The Limits of the Generalist

I am a Diplomate of the American College of Laboratory Animal Medicine (ACLAM). That means I am an officially designated board-certified specialist in the field of laboratory animal medicine and can use that designation in describing/portraying/marketing my professional credentials to the public without incurring the wrath of the AVMA and its lawyers. To become board-certified in laboratory animal medicine requires several steps. Thirty years ago, these included at least 2 years of practicing in this field under the tutelage of someone already board-certified by ACLAM (or you could go the "practical experience" route with at least 6 years on your own) and being first author on a scientific paper published in a peer-reviewed journal. After I graduated from veterinary school in 1982, I was fortunate to get an appointment as a Post-doctoral Associate[1] in the Division of Comparative Medicine at the Massachusetts Institute of Technology, under the direction of Dr. Jim Fox. A detailed description of one's training or practical experience and scientific paper were submitted in a lengthy application to the ACLAM Credentials Committee that determined if you qualified to sit for an all-day written examination along with everyone else who was trying to get board-certified at the same time. The examination covered husbandry, normal biology, and natural diseases of species commonly (or not!) used in research and testing, vaccines and medicines for treating their diseases, how these species were used to model human illnesses and injuries, and applicable laws and regulations. If you had passed the examination, you were welcomed into ACLAM as a board-certified specialist (Diplomate).

There are numerous other AVMA-recognized specialties within the veterinary profession, comprising many of the same fields found in human health care, such as ophthalmology, radiology, dermatology, and dentistry, to name a few. There are also recognized veterinary specialties that don't have physician counterparts, such as poultry or zoo medicine. The American College of Veterinary Internal Medicine is the largest at the time of this writing, with over 2600 Diplomates. The American College of Veterinary Pathology is the oldest, extending back to 1949. One of the newest is the American College of Animal Welfare, recognized by the AVMA in 2012. Close to 12,000 veterinarians are board-certified under the AVMA, some in more than one specialty.[2] Other countries or regions may have their own counterpart specialty organizations. My field alone includes the European, Japanese, and Korean Colleges of Laboratory Animal Medicine, each with their own bylaws and credentialing requirements.

Notes in the Category of C. https://doi.org/10.1016/B978-0-12-805070-5.00007-2

Truth be told, laboratory animal veterinarians are more like generalists than specialists when compared with other veterinary specialties. We often engage a wider spectrum of medicine and surgery than most of our peers, differing only in the sense that the animals to which we tend are used in research, testing, and education instead of companionship, food production, sport, exhibition, conservation, etc. Any given day may present a laboratory animal veterinarian with managing an infection that threatens the health of thousands of rats or mice in a colony (thus drawing on one's knowledge of epidemiology and microbiology), assessing a new research protocol for optimal pain management regimens (pharmacology, anesthesiology), training researchers on making and closing skin incisions properly (surgery), and advising an investigator about managing animals with specific metabolic conditions (internal medicine, clinical pathology).

Large or complex veterinary institutions in non-laboratory animal sectors, such as multi-specialty clinics and academic teaching hospitals, employ more than one category of specialists for treating sick or injured animals. Yet laboratory animal care programs typically have only laboratory animal veterinarians on their professional staff, occasionally accompanied by one or more veterinary pathologists to manage an in-house diagnostic service. Some of these laboratory animal veterinarians may be board-certified in a second (and third!) specialty, but they are exceptions to the rule. Hence, the question arises: could we serve researchers and laboratory animals better if our programs were more varied in their professional staffing? Rather than having six or seven laboratory animal veterinarians, all of whom have the same generalist training for species commonly used in the laboratory, why not also employ veterinary specialists with greater depth in complementary areas?

For example, animal imaging has undergone profound growth over the past 10 years. Sophisticated scanners relying on magnetic resonance imaging, computed tomography, positron emission tomography, and more are used to monitor if and how certain cancers grow or metastasize, how damaged spinal cords heal, how heart defects evolve, how cerebral strokes behave, etc. in mice and other small species. This has allowed scientists to track what's going on inside individual animals with greater precision and over longer periods of time. This also often reduces the number of mice needed for a given experiment because one doesn't have to euthanize groups of them at various time points in order to evaluate their organs post-mortem. Larger animals, such as monkeys and pigs, are commonly used in combination with brain scans and cardiovascular imaging, respectively, as well as for many other protocols. So I've wondered why aren't we recruiting veterinary radiologists to assist in further developing these models? The imaging physics and chemistry are the same, and the subjects are all four legged (except for monkeys). Just the purpose of the animal is different. It makes sense that a veterinary imaging specialist could provide helpful expertise in positioning the animal subject, determining optimal tissue penetration energies, and so on. That seems more efficient than relying on the physicians or scientists to figure it out on their own, perhaps with the assistance of a laboratory animal veterinarian who likely doesn't have the depth of knowledge and experience of a board-certified veterinary radiologist.

Another example involves nutrition. Laboratory animals are usually fed a standardized diet, primarily to avoid unwanted variations in nutrient levels and quality that could undermine the reliability of the data generated from these animals. And it's common to provide diets that match specific animal needs such as breeding and lactation versus diets for normal physiological maintenance. In addition, species of higher intelligence, such as monkeys and pigs, are offered a wider variety of food to help minimize boredom. But what's routinely missing from our animal food cupboards are meals designed for *restorative* nutrition, i.e., specific nutrients combined to ameliorate metabolic diseases and other chronic conditions. The pet food sector has been well represented for decades in this regard with companies offering special rations targeting conditions such as congestive heart failure, liver failure, kidney failure, sensitive gastrointestinal tracts, and more recently, obesity and osteoarthritis. Rather than offer laboratory animals diets designed only for normal healthy states, why not use counterparts to those pet medical diets, provided there's no conflict with the scientific aims of the experiment in question? This could not only reduce disease severity but also possibly extend the period of time during which these animals are available for study before they die. It also could better mirror the human patient's entire spectrum of treatment by mimicking how human patients are managed nutritionally, which in turn may generate more relevant animal model data in pursuit of medical cures. Thus, there might be great value inviting Diplomates of the American College of Veterinary Clinical Nutrition into our world, to advise on dietary support for a wide variety of disease conditions in laboratory animal species, most notably mice, rats, monkeys, and pigs.

A third example of how other veterinary fields could collaborate with laboratory animal medicine may be less obvious since it involves program management rather than clinical expertise. This particular relationship first occurred to me during an animal welfare conference named "The Conversation," convened in November, 2013 by the AVMA.[3] Billed as "an intraprofessional conversation about animal welfare," this 2-day meeting invited veterinary specialists, animal welfare scientists, and bioethicists to discuss how the veterinary profession could further improve animal welfare across the various fields of practice (e.g., companion animal medicine, poultry medicine, equine medicine, livestock medicine, zoo animal medicine, wildlife medicine), via whatever organized support AVMA could provide. I was honored to be asked to deliver remarks on behalf of laboratory animal medicine. During the lunch break on the first day, I happened to sit with a couple of fellow laboratory animal veterinarians and several pet shelter practitioners. One may expect we would be blood enemies of a sort, since my field uses animals for experimentation and testing, whereas theirs is dedicated to improving the health of abandoned or neglected animals in hopes of finding them a good home. But perhaps given the intent of the symposium to promote a profession of shared values or because it really didn't matter, we easily initiated a friendly conversation. As we introduced ourselves and our affiliations around the table, it quickly became apparent that our respective interests had much in common: large numbers of animals confined in close proximity, some of which had minor to serious diseases or injuries (in the vivarium, almost always induced for scientific purposes, whereas in the shelter, never); constant turnover of occupants; expensive facilities with high fixed costs; dependence on compassionate and competent animal care staff; significant adverse publicity when things go wrong (or even when they don't). It was also apparent from that meeting and others I've had since with senior shelter veterinarians that laboratory animal programs are usually more wealthy and better equipped than pet shelters. One reason is because vivaria are highly regulated and therefore require expensive infrastructure for compliance while shelters depend on often unpredictable charitable donations for much of their funding. Similar to the first two examples above, we should be asking what do shelter veterinarians know that may benefit us? What have shelter veterinarians learned about colony health management on a shoestring budget that we could adopt for managing our programs in a more cost-effective fashion? To that end, how could both parties engage in comparative management conversations and host reciprocal site visits to familiarize each other with their respective management strategies while sharing tips and avoiding wrong decisions? While not an AVMA-recognized specialty, the Association of Shelter Veterinarians boasts over 1500 members.[4] Therefore, it shouldn't be hard to find an approachable veterinarian willing to answer these questions. One approach that could catalyze such discussions and would be easy to launch is to offer to donate unwanted or unusable laboratory animal equipment and supplies to a local pet shelter. Many laboratory animal programs occasionally have equipment or supplies that are no longer needed because the species for which they were acquired is gone or the animal model has been abandoned.[5] Nothing usually works as well as free stuff with no strings attached for initiating a conversation. And helping pet shelters with the noble work they do should also make everyone in the vivarium feel better about their own program.

It's easy to imagine other animal care and use needs served by specialists in other veterinary fields. And such arrangements don't have to involve full-time employment. Laboratory animal care programs could retain other veterinarians as part-time employees or consultants, and academic institutions could arrange for sabbaticals by visiting veterinary specialists. Metropolitan areas with multiple biomedical research institutions could establish consortia that offered other veterinary specialists enough cases or models in their field to attract their interest and share the costs. Other anticipated benefits from these interactions include advances in other fields from what's learned in the vivarium and a better understanding of laboratory animal medicine by the other veterinarians. So let's be more imaginative and try harder. After all, the veterinary perspective around "One Health" applies beyond humans to other animals in other settings.

# Chapter 8

# From Bedside to Cageside

The phrases "comparative medicine" and "translational medicine" have been commonplace in animal research for many years. Used to encompass the generation and transfer of new knowledge about disease and injury, as well as new treatments, these categories of investigation are uniformly applied in only one direction, i.e., starting with animal models and ending with patient care. But there's no reason why such knowledge couldn't flow in the opposite direction for the betterment of laboratory animals themselves.[1] Certainly, there are elements of biology, safety, practicality, and economics that need to be assessed to determine what's applicable and what's not. But there is an abundance of possibilities, both well established and emerging, that could prove fruitful if we only engage our imagination. Examples in several major health care categories will be considered below.

## POINT-OF-CARE DEVICES

So-called because they are designed for use outside the conventional hospital setting, such as the doctor's office or in the patient's home, point-of-care (POC) technologies are intended to offer quick, simple, and inexpensive means of diagnosing and monitoring a patient's status. A couple of old standbys in this regard are the urine test strip that immediately detects unusual levels of various excreted substances, and blood glucose monitors routinely used by diabetics. More recently, the use of pulse oximeters has become widespread for monitoring blood oxygenation levels via a small finger clasp. Contrast that to the old approach of drawing an arterial blood sample and rushing it to a sophisticated machine in an advanced laboratory. The common thread here involves diagnostic devices and assays that don't require someone to process, package, and ship biological samples to a distant laboratory (even if it's in the same building). Instead, results are available immediately and clinical decisions can be made sooner. These and other POC devices are widely used on and for laboratory animals.[2] But there are other examples in development or already on the market that could be employed in the vivarium now or sometime soon. Some POC assays will involve smart phones for even greater convenience and speed.[3] Why not employ these at "cage-side" so that laboratory animals can remain in familiar surroundings and not have to be wrestled and stressed from collecting the sample or being transported to wherever the current machine resides?

Notes in the Category of C. https://doi.org/10.1016/B978-0-12-805070-5.00008-4

One possible scenario could involve systemic infection and inflammation, either induced or unintended. A prototype microfluidics device has been described that can detect subtle changes in neutrophil behavior in the blood of healthy versus septic mice. When neutrophils are activated by infection or inflammation, they can migrate in different numbers, at different speeds, and in different directions.[4] If this technology can be simplified for use within the vivarium, then a small drop of blood obtained cageside could be quickly analyzed to determine if sepsis or large-scale inflammation is worsening or improving. And because the volume of blood needed would be small, more frequent sampling can be performed to track progression of the problem with greater accuracy. That in turn may enable either more humane endpoints (e.g., employing a change in circulating leucocyte behavior rather than requiring the animal's health to decline even further) or better clinical care, such as changing the dose or type of antibiotic used to treat an iatrogenic infection, especially if the original bug wasn't identified.

Speaking of blood sampling, a common concern in laboratory animal medicine involves animals that are subjected to repeated blood collection. IACUC policies and protocols usually impose limits on how much blood can be taken over a given period of time in order to avoid anemia. But even if the animal is noticeably weak or pale (the latter being harder to detect in an albino), it's still not easy to determine if these limits are being exceeded or if these limits need to be reduced for that specific animal's benefit. To that end, envision POC devices in the future that will measure hematocrits and perhaps even characterize erythrocytes for size and iron content, as conveniently as pulse oximeters measure blood oxygenation levels today. Cageside monitoring of this sort for anemia could be helpful not only for research purposes but also to reveal excessive blood withdrawal beyond what's permitted under that protocol.

## PRECISION MEDICINE

At least 20 genes have been shown to influence the metabolism of at least 7% of all FDA–approved drugs in humans, with a growing call to screen individual patients to determine how well they will metabolize these drugs.[5] There are now genetic tests to detect similar differential drug metabolism capacities in various dog breeds for veterinary decisions.[6] Why not apply the same approach for other species that just happen to be in laboratories instead of just for pets in homes?

One area that offers progress in this regard is analgesia. A landmark paper published almost a decade ago showed that both rats and humans had the same variance in pain tolerance due to the same versions of a single gene.[7] Others have since provided more evidence of the relationship between pain sensitivity and genetics for human patients, especially after surgery.[8] One can easily envision genetic screening of laboratory animals at the time of their assignment to various treatment groups in Category D protocols to see which animals may require more or less pain medication, as a means to optimize drug efficacy and avoid toxicity. And even better, such testing can be population-based for inbred animals because of their genetic homogeneity. As long as these genotypes are conserved, their affiliated pain sensitivities and analgesic drug responses should remain true and reliable. I'm not suggesting that genotypes associated with greater pain tolerance should be necessarily favored over those with less pain tolerance if that favoritism discourages important research; only that for experiments involving pain that require animals that have a genetically based lower pain tolerance, analgesic dosing regimens should be increased and animals monitored accordingly for pain. It's ironic that inbred mice have been "humanized" to study the genetic basis of differential drug metabolism in man but no one has employed this approach to create genome-based pharmacological response profiles to apply solely in a clinical fashion to various mouse lines themselves, especially for analgesics and highly toxic drugs.[9]

(Another intriguing sidebar: have inbred mice of both sexes ever been used in "humanized" modeling to predict how sex may affect drug metabolism in women versus men? In a related vein, have humanized mouse models of drug metabolism specifically used neonatal and adolescent mice as models for developmentally matched humans? If not, then there's lots still to do and possibly much to be gained by expanding humanizing technologies in inbred and genetically modified mice.)

For outbred laboratory animals, another application where precision medicine may provide benefit is in the field of experimental organ transplantation. Acute organ rejection, immunosuppressive drug toxicities, and eventual failure of transplanted organs to thrive are all risks to both human recipients and their corresponding animal models. Serial biopsy of the new organ for histological analysis has been the gold standard to determine the nature of the problem. But for more than a decade, it's been known that genomic variation of both the organ and recipient can be clinically relevant even when biopsies yield identical lesions or infiltrates when examined under the microscope.[10] When I was involved with veterinary support to large animal organ transplant models some time ago, an occasional frustration was varying levels of cyclosporine in the bloodstream in and between animal subjects, be they NHPs or pigs. Trying to titer the correct dosage, both to attain therapeutic levels of circulating drug and to keep the levels consistent within a given treatment group per experimental design, was a problem. If cyclosporine levels were too low, the organ could be rejected; if these levels were too high, serious side effects could jeopardize the health of the animal recipient even more. But what if consistency of circulating cyclosporine between animals was less important than establishing the optimum balance for each individual animal? Perhaps a simple genomic assessment pre-surgery could have shown which version(s) of which gene(s) of a given organ recipient as well as those of the organ donor would predict the response to cyclosporine. That in turn may have spared some animals and also strengthened the statistical power of the experiment's results.

## MULTI-CENTER TRIALS

New veterinary knowledge is almost always reported from individual laboratory animal care programs and then disseminated broadly, even though the origin of that knowledge was a single source and the results not formally validated by organized repetition prior to publication. This practice is reflective of centuries of individual scientists toiling alone, perhaps aided by one or two assistants. Contrast that with today's research laboratories, including those engaged in comparative medicine, that are large and have highly organized groups of investigators at senior and junior levels. In addition, scientific papers are more multi-authored than ever before and often representing more departments and institutions than ever before.[11] These multiplicities are a consequence of many factors and speak to the growing fluidity of collaborations in which teams of researchers may assemble to tackle a particular question or share a rare and expensive resource, and then disperse after the experiments are performed.

Now consider the strong tradition of performing multi-site clinical trials to determine the safety and efficacy of new medical products and procedures before they can be approved for marketing by the FDA and other health regulatory agencies and their costs reimbursed by insurance firms. Such trials are deemed beneficial for several reasons. First, they can enroll a larger and more diverse population of patients than any single hospital could attract or accommodate in a given period of time. Second, they may correct for any single physician bias, institutional or geographical peculiarity, or other variable that may reside at any one site and skew the results. Third, they enlist a wider swath of practitioners whose cumulative experience and knowledge may better inform the trial during its conduct and data analysis.[12]

Would laboratory animal medicine and its stakeholders similarly be better served by following up initial research reports with a broader assessment coordinated amongst multiple programs? Given the recent crisis in irreproducibility of scientific data, including but not limited to biomedical research and animal models, this option may be timelier than ever before.[13] There is legitimate concern about the influence on study data variation by differing husbandry components such as the type of cage, bedding, environmental enrichment materials and practices, food and water, and even animal care personnel. In addition, we well know that different genetic backgrounds, even amongst so-called outbred groups, can generate different results. Today's answer to all of this is almost always individual reports popping up one at a time, with the Discussion section of the paper including the obligatory acknowledgement that the results obtained may or may not occur under other circumstances involving [fill in the blank]. By contrast, would a coordinated multi-site approach identify which, if any, variables may lead to divergent data faster and more effectively, to the benefit of our customers and their animals?

Such stratagems wouldn't have to involve complicated or expensive biomedical science or sophisticated animal models to be useful. They could also address simpler questions involving husbandry or common veterinary practices. In either event, one could imagine employing a classic multi-site trial design in which standard "enrollment" criteria are applied uniformly, perhaps even assigning animals across all participating programs to different treatment groups via a formal randomization process. Studies could also be double-blinded, where pertinent, so those administering treatments and collecting response data would not know the particular drug or dose they were administering. But unlike official clinical trials, study design and results would not be regulated or submitted to a governmental agency for review but rather shared via conventional channels.[14] And using the clinical trials paradigm as a model doesn't mean a rigid approach to multi-site study design; there's a growing recognition that human clinical trials can be more flexible in their planning and execution without jeopardizing their validity.[15] Along these same lines, clinical multi-site trials sponsored by NIH will soon require review by only a single institutional review board (IRB) rather than multiple IRBs as had been the practice.[16] Perhaps if multi-site IACUC protocols become more common, similar regulatory relief will be provided to participating institutions so that only a single IACUC can represent all the others involved.

New laboratory animal medicine practices or products aren't the only targets that could benefit from coordinated multi-site experiments. Previously published findings from a single source that remain poorly adopted because of skepticism or are under-appreciated in terms of their potential impact could also benefit all of us with a second and expanded look. Coordinated multi-site experiments could bring more publicity if they confirm original findings from a single laboratory or generate healthy discourse if they don't. For example, antibiotics provided in drinking water bottles to laboratory rodents for the prevention of post-surgical infections remains a common practice, at least amongst researchers if not veterinarians. That's despite the fact that this drug regimen has repeatedly failed to either prevent bacterial infection or achieve inhibitory drug levels in the bloodstream of mice.[17] Would a prospective multi-site trial, even a modest one that merely tracked evidence of infection in dermal incisions on active protocols or measured circulating levels of the drug in the blood, and not involve intentional inoculation of surgical wounds with bacteria, be useful in convincing others about the futility of this practice? On the other hand, if contrary results arose in the course of such a multi-site study, then the details could be dissected in a cooperative and rigorous manner and advance the field just as well as a uniformly negative conclusion.

In addition to medical treatments for individual animals, consider multi-site trials for better colony health outcomes such as bioexclusion practices against rodent viruses. Currently, there are very many opinions about very few ways for excluding unwanted infectious agents from barrier rodent colonies. Most of these approaches are based on antiquated strategies from decades back when most rodent cages couldn't limit the spread of microbes and when most rodent colonies were infected with a myriad of downright nasty pathogens (vs. the quasi-commensals with which we are mostly concerned about today). In the old days, showering in or out, combined with or without a complete change of scrubs, made sense when going from one facility, one colony, or one room to the next because personnel were a logical fomite for spreading around viruses, bacteria, and parasites. Since the advent of micro-isolation cages and accompanying cage-changing stations that provide chambers surrounded by streaming curtains of purified air, we've had 30 years of peer-reviewed scientific publications confirming that such cages faithfully isolate their contents, including microbes, without the risk of spreading if used correctly.[18] That not-so-recent change in how barrier rodent colonies are equipped should, in turn, have liberated programs from obsolete and wasteful approaches to colony biosecurity with respect to facility design and traffic patterns (e.g., separate clean and dirty corridors for soiled vs. clean caging), adequate personal protective equipment, and other elements. Rather than be a lone voice in the wilderness with another peer-reviewed paper that (yet again!) confirms the biosecurity insurance and assurance of a proper micro-isolation caging technique,[19] would a coordinated multi-program assessment be more convincing and lead to progress faster?

What if someone wanted to try a more radical approach to avoiding and detecting rogue pathogens in an SPF colony that could save even more time or money? How would one go about testing its performance versus current practices? Employing a rodent virus for this purpose is out of the question even if it's not currently labeled as a pathogen in that facility. That's because it may spread beyond the confines of the experiment and could influence subtle endpoints in someone else's experiment in a way that wasn't predictable. Because of those legitimate and real concerns, I've always been intrigued about using plant viruses for this purpose. Not only would they not infect animals, including us, but they could be selected to mimic a particular rodent virus of interest based on shared physical properties.

Take, for example, mouse parvovirus (MPV), a vexing colony problem to say the least. Since it is non-enveloped, it can survive seemingly forever in minute amounts of organic debris anywhere in a facility and then reappear in a colony even after expensive and exhaustive eradication campaigns. Cucumber mosaic virus is an attractive candidate for this purpose. It is similar in size (29 nm diameter) and shape (icosahedral capsid) to MPV and also non-enveloped. Plus it's a popular laboratory virus so preparing aliquots for dissemination and detecting remnant virions should be easily performed by a collaborating plant pathology laboratory.[20] It's a ubiquitous agent in agricultural and economic produce settings (i.e., we and other animals are exposed to it all the time). Furthermore, current evidence indicates that its presence in insect vectors is purely mechanical without infecting the bodies of those arthropod vectors. So any threat of infecting or influencing vertebrates is remote, at best[21]; even less so where contemporary micro-isolation rodent caging is employed. Thus, there should be no special biocontainment precautions necessary or any risks to colony animals, even though its use in a vivarium may require Institutional Biosafety Committee review and approval in advance.

(One alternative to using a non-animal virus would be to rely on a fluorescent powder as a tracking material to determine which bioexclusion practices were reliable and which needed to be improved, similar to an experiment in which I served as the guinea pig technician some 40 years ago. As the central component of the study, I was frequently sampled for powder residue while traipsing around a vivarium and feeding special agar-based diet to rats.[22] But that approach would not involve a self-replicating organism, such as a virus, that could be assayed for an increase in numbers or locations as well as its mere physical presence and, therefore, would not be as convincing.)

Rather than measure residual cucumber mosaic virus levels after varying rodent pathogen control measures in a single facility, a coordinated multi-site trial could be more informative and more applicable to the laboratory animal medicine field as a whole. That's because programs may differ a little in their biosecurity practices. So different environments could be assessed collectively to see which prevention or eradication variables may be significant or inconsequential. Admittedly, we're not fulfilling Koch's postulates here with the actual virus of interest. But since MPV is a recurring concern precisely because of its environmental persistence (as a consequence of its physical structure and size), modeling these exact parameters in a plant virus with similar features should suffice to confirm or refute the dependability of a particular biosecurity scheme for colony protection. And if a multi-site trial offers a more effective and less costly alternative, all the better for the rest of us.

## CLINICAL NUTRITION

In-patient care, by necessity, includes dietary considerations at mealtimes. What is the patient's metabolic and physical status? How do those elements dictate what sustenance of a non-pharmaceutical nature is to be offered, in what physical form and schedule, and for how long before a change in health status triggers a change to a different diet? The medical field of clinical nutrition is devoted to answering these questions, for both human and veterinary patients, with board specialties and other credentials established for each profession. From eating normal meals to total parenteral nutrition, the options are almost endless and must be designed with the patient's immediate and evolving needs in mind. While we now offer NHPs a veritable cafeteria of food options for environmental enrichment purposes, the literature and practice of adjusting nutrient intake based on a laboratory animal's *clinical* status is scant by comparison. As evidence, only three peer-reviewed articles involving a clinical nutrition intervention to ameliorate the severity of induced illness or injury were identified over a recent 10-year span (2007–16) in three journals where such papers would likely appear: Comparative Medicine, Journal of the American Association for Laboratory Animal Science, or Laboratory Animals (UK).[23] Even if there's an extensive body of work for laboratory animals buried within another discipline, these three journals are where the subject matter should be at least publicized for sharing amongst laboratory animal practitioners.

Getting more specific, what about special diets for kidney and liver transplant recipients, especially if their entire native organs were removed prior to surgery in order to evaluate the new organ more accurately? Would an animal that is serving as a liver failure model not only feel better but also generate more research data over a longer period of time if fed a low nitrogen/low protein diet (provided, of course, that hyperammonemia or resultant hepatic encephalopathy wasn't the focus of the study)? Along the same line of reasoning, why not offer low-salt diets to animals modeling congestive heart failure? How should an omnivore species, such as a laboratory dog or pig, be managed in a clinical nutrition mode differently from a carnivore, such as a laboratory cat, or an herbivore, starting with rodents? What would a "bland" diet look like for a mouse with inflamed intestines that is being used as a model for Crohn's disease? Are there diets that would be less nauseating for animals that are on cancer protocols and administered conventional chemotherapy drugs or given opiates for post-surgical analgesia? Speaking of surgery, an animal's energy needs are significantly higher after a major surgery due to greater demands from tissue healing and repair. Thus, would caloric supplementation be medically helpful, and when should extra calories be provided if at all?

To that last question, it has become common in some hospitals to provide carbohydrate-containing drinks (CCDs) to human patients before their surgeries to promote recovery afterward even though evidence of quantifiable benefits of this practice has been mixed.[24] Yet one study in rats showed access to a carbohydrate drink rather than fasting prior to surgery resulted in a reduction of bacteria in vital organs and mesenteric lymph nodes after experimental ischemia-reperfusion in terminal abdominal surgery.[25] Are these and other studies sufficiently compelling to encourage the provision of CCDs to laboratory animals in general before surgery, especially for small animals with much higher basal metabolic rates? If not, what other parameters need to be investigated?

(If your opinion is influenced by a fixation on overnight fasting of laboratory animals prior to general anesthesia in combination with surgery, there is ample evidence that CCDs can be administered to human patients safely up until two hours prior to induction of anesthesia with no increased risk of aspiration. Hence, the recent Canadian policy assessment referenced in the paragraph above concluded that "the common practice of NPO from midnight onwards is unnecessary" for human surgery. Furthermore, laboratory rodents fasted before challenge with a stressful situation such as trauma (hemorrhage) or infection (endotoxin) were reported to exhibit reduced muscle and cardiac function, decreased immunologic performance, and decreased survival rates, among other adverse outcomes, in comparison to fed animals.[26] Thus not only is routine fasting unnecessary but also may be detrimental, depending on the species and circumstances.)

Beyond the clinical diet itself, there's also room for improvement in monitoring food intake as well as the age of any food that's not eaten. Why don't we have so-called "smart" food hoppers that can track food levels over time and also alert staff when it's time to change the food so that it won't spoil? The core technology is certainly available, and an entire vivarium doesn't have to be outfitted—just cages for animals at the greatest medical risk of inanition. For starters, how about a special wire bar lid for rodent cages that can be easily swapped with the regular kind but is equipped with weight transducers that transmit food loads along with the date and time since they were last loaded? Smart phones could be linked via an app to sensor outputs via local wireless networks and tracked from anywhere at any time by any one with the requisite security access. Sensors also might be sensitive enough to be "disturbed" whenever food pellets are being dislodged by cage inhabitants in the course of eating. That in turn would be a nice bit of information when tracking the eating patterns of normal versus sick animals. One multi-national rodent cage vendor has already started down this road[27]; additional players are welcome.

In closing out this chapter, there are lots of exciting moves afoot to improve the effectiveness of health care to patients in the United States and elsewhere. Part of this is a welcome focus on immediacy and efficiency, either at bedside, in the physician's office, or through a home health care attendant. There's no reason why these developments, as well as more established practices, shouldn't be considered for the laboratory animal patient and adopted if possible.

Chapter 9

# Laboratory Animal Psychiatry

During my initial years as a laboratory animal veterinarian, as I started to attend national meetings and other gatherings of our tribe, the topic of animals suffering repeatedly caught my attention. The subject was hard to ignore since it was constantly raised by those opposed to the use of animals in research and testing. My more wizened colleagues advised me off the record to espouse and disseminate the position that only humans can "suffer" while non-humans cannot. That was because non-human animals don't have the same cognitive capability for "feelings" that we do, supporting the notion of the singular human capacity for suffering. By contrast, as the semi-official party line went, non-human animals were only capable of experiencing "distress," a label deemed a more primitive state shared across more species. Thus when speaking publicly or to each other, we were discouraged from talking about animals' feelings and from describing those animals as suffering if the animals appeared to be in pain or were exhibiting pathologic behaviors. Saying otherwise would inaccurately humanize (literally) the animals' plight and give opponents of animal research more of a foothold to play the sympathy card.

Even while I dutifully followed such advice, I was torn by evidence to the contrary, as both a veterinarian and a pet owner. My family has frequently had dogs as central members of our household. As they got old and their health failed, it was obvious that their demeanor, activity level, appetite, attitude, etc. were perfectly consistent with a state of suffering, irrespective of any pain medication they may have been given. In a more professional and objective vein, I also had trouble reconciling the established utility of rats and other species as long-standing models of chronic depression, anxiety, and fear in human patients with the conventional wisdom that these animals couldn't feel (suffer) in a fashion identical to those patients. If comparative behavioral science was such as an established research discipline and these animal models were regarded as the gold standards for studying human psychiatric conditions, how could the respective emotional states of other animals and us be so different?[1]

I found myself in the middle of these semantic conflicts when I accepted an invitation in 2006 to serve on an ILAR panel appointed by the National Research Council of the National Academy of Sciences. Our assignment was to issue a consensus expert report on detecting and alleviating distress in laboratory animals, as an update to an ILAR report published 14 years earlier that also included pain.[2] The Academy and the new report's sponsors wanted to divide the subject matter into two documents, ours devoted to distress while another

Notes in the Category of C. https://doi.org/10.1016/B978-0-12-805070-5.00009-6

group was to update a separate pain guidance document. There were lots of new findings in both fields, and distress was gaining more attention as an animal welfare concern in its own right. The composition of our panel included veterinarians, physicians, animal welfare specialists, animal behaviorists, physiologists, and neuroscientists, arriving from Honolulu, London, and points in between. We were to convene twice in person while also communicating frequently via e-mail. Our charge included that "Specific emphasis will be placed on the identification of humane endpoints in situations of distress and principles for minimizing distress in laboratory animals," to inform IACUCs of possible alternative outcomes in their reviews of protocols that involved distress.[3]

Because the scientific discipline of distress was so broad, a sensible place for us to start was to define distress so that IACUCs would be advised about the evidence-based options for ameliorating any distress integral to an experiment and identify appropriate veterinary interventions in the vivarium. However, it was clear early on that we would have difficulty reaching consensus on what was to be included or not in our definition. Some panelists respectfully contended that we should confine our attention only to *physiological* distress because it was easily measurable and long-standing in the scientific literature. Tracking vital signs and blood glucocorticoid levels provided an established framework within which physiological distress had been studied for decades. In addition, failure of an organism's homeostatic capacity to return to physiological normality after a stressful experience was a classic threshold definition for distress and relatively easy to gauge; anything else was too speculative. Other panelists asserted just as respectfully that *emotional* distress should be incorporated into the issued report because it was just as relevant as physiological distress with respect to animal welfare, in spite of the fact it could neither be measured by blood tests or other simple diagnostic assays nor was it likely to be experienced by animals in exactly the same way as humans.

We muddled through this debate and our assignment, and in the end declined to issue an all-encompassing definition of distress. Instead, we acknowledged that much was still unknown about distress and recommended that more research be performed to better understand how distress of all kinds could be detected and avoided or minimized. Quoting from our summary statement, "Scientific research does not yet support objective criteria or principles with which to qualify distress, objective scientific assessment of subjective emotional states cannot be made, and while there is often a measure of agreement on the interpretation of physiologic and/or behavioral variables as indicators of stress, distress, or welfare status, there is not always a direct link." So while we could not establish a consensus definition of distress for laboratory animals, we felt confident enough as a group about where to draw the line in order for the above statement to remain in the official version of our report. In the meantime, we encouraged fealty to the 3Rs and to use one's heart as well as one's head to identify and alleviate distress in laboratory animals whenever possible.[4] In one sense, we did well enough—the sponsors were satisfied with the report and laboratory animal distress finally received singular recognition as a worthy concern separate from pain. But in hindsight, I wonder if we missed the boat.

Ignored, or at least not adequately appreciated, during the time our panel was doing its work was the existence of a body of compelling scientific evidence that humans and many common species of laboratory animals share the same part of the brain for expression and control of emotions, and it wasn't the frontal cortex, the area of advanced (human) cognition. Instead, anger, fear, sadness, sex drive, and other emotive states are centered in the limbic system. This region is separate from areas of "higher" dispassionate thinking and conserved across many laboratory mammals. In addition to shared anatomy, elements of the limbic system display the same kind of physiological activation when the organism is exposed to the same types of stressors. In the course of our panel's discussions, I became aware of the work of Joe Ledoux and his colleagues at New York University, where his laboratory was investigating how the amygdala was the central traffic interchange in the brain for emotion signaling, regardless of whether the organism was a rat or a person.[5] I later heard an interview with neuroscientist, Jaak Panksepp, on National Public Radio. Panksepp was a media draw at that time because he had published research claiming that laboratory rats were ticklish, audibly "laughed" when tickled, and would voluntarily come back for more.[6] I then checked out Panksepp's oeuvre and learned that he was widely published on the science of animal emotions and how similar those emotions are to ours.

These and other researchers' published findings introduced me to the field of affective neuroscience and presented a new way to think about animal emotions and animal suffering. In one of Panksepp's books, also part autobiography, he recounts how he initially raised the same questions as mine about the capacity of animals to suffer in human-like terms, and how he had been similarly advised not to approximate the two.[7] But Panksepp, Ledoux, and others demonstrated that emotional states are more primitive than conscious thoughts and (as a consequence) are more shared between humans and non-humans than previously accepted. If my understanding is correct, animal models of basic human emotional disturbances are scientifically reliable precisely because the animal's emotional processes are close enough to ours to offer reliable scientific insights. Hence, the intellectual disconnect between suffering of (laboratory) animals and animal models of suffering could be reconciled.[8]

After I learned about the evidence-based legitimacy of claiming that, yes, animals actually suffer anatomically and physiologically like we do, I began to think differently about laboratory animals with various behavioral pathologies that arise spontaneously versus those that are intended outcomes of experimentation. If such behavioral problems in animals medically resemble human mental illnesses, could human psychiatry and related fields offer a more accurate means of diagnosis and better therapies for these animals? I'm not suggesting anything as daft as Freudian psychoanalysis through insightful conversations with an animal patient (unless one happens to be Dr. Doolittle). Other effective treatment modalities are available to laboratory animal medicine and some are being used even though they're not (yet) labeled "psychiatric." The question posed in this chapter is could we practice better veterinary medicine on laboratory animals that present with unanticipated mental illness if we managed those subjects as mentally ill patients?

Let's start with stereotypy, a particularly troublesome behavioral problem in captive animals and characterized by continuously repetitive movements that appear to have no purpose. This condition is more likely to occur when animals are maintained in small and barren enclosures and can be observed in zoos as well as vivaria. Think about elephants or large cats pacing back and forth along the front of their cages, or monkeys continuously turning somersaults, sometimes for hours at a time. Stereotypy is also common in laboratory monkeys, possibly a consequence of their higher intelligence, space limitations, or lack of stimulation that comes with conventional primate caging, and early life experiences in captivity.[9] Even though it's not a common condition, it can be very frustrating to treat and resolve. The usual sequence of intervention starts with determining if there's an obvious origin of the behavior. If the monkey is responding to a continuous anxiety-generating stimulus without a coping option available, such as being housed directly across from another NHP that for some reason is visually intimidating, then one can change the location or orientation of the affected animal's cage so that the offending stimulus is avoided (while being careful not to initiate a new conflict of proximity or visual access that bothers other animals in the room). Or if you think the monkey has gotten into a behavioral rut from too much of the same routine, then more variety in terms of food, toys, music, or videos is worth trying. Occasionally, none of this works. And if the animal is in a critical phase of its experiment, other tactics, such as drugs, may be tried to kick start it out of its stereotypic state. But sometimes a pharmaceutical option is not compatible with the protocol so the animal may either remain in a distressed state until the experiment is completed, or removed from the study before it's completed.

We had an opportunity to depict and study simian stereotypy in a novel way years ago after learning about a company that had developed a fancy high-tech vest that measured multiple posture and activity dimensions plus physiological parameters such as heart rate, electrocardiogram, respiration rate, and thoracic versus abdominal breathing.[10] Designed for humans engaged in strenuous physical activity, this vest was being marketed to elite athletes, fire fighters, and others working under physically demanding conditions. The wearer of this vest could easily see how an increase or decrease in physical fitness would immediately be captured and displayed for smarter and more effective training. This company was also targeting sleep clinics and patients with sleep disorders. Wearing the vest to measure numerous biological phenomena while sleeping was attractive because no cumbersome wires or leads had to be connected from the subject to a machine; all the requisite technology was contained in the vest. I was intrigued about having monkeys wear these vests to study stereotypy and other situations involving motion or vital signs, and was delighted to learn from the company that they were developing animal-sized models.[11] What we discovered by employing the vest was that some macaques with stereotypy had dramatically shortened periods of rest or sleep, as illustrated in Figs. 1 and 2.[12] These printouts were helpful in more accurately assessing the efficacy of various veterinary interventions to reduce or eliminate stereotypy in individual cases.

(Other animals besides monkeys exhibit repeated behavior that appears excessive to us but may not be stereotypy. For example, some mice will chew but not swallow most of their food, leaving the floor of the cage covered in powdered chow. The mice otherwise appear normal; however, this drives the animal care staff crazy in having to replenish the food hoppers and change cages much more frequently than planned. But excessive chewing may reflect an unsatisfied nutritional need rather than a drive to handle and gnaw food in a stereotypic fashion.[13] An even more common activity in mice without apparent purpose is barbering [chewing or pulling the hair off the face and head of cage mates]. Barbering was believed to be a consequence of social dominance back in the day when I was a post-doc, in which the only mouse in a given cage with all its facial hair was believed to "barber" other mice that were lower in the cage's hierarchy. However, it's been since shown that barbering is an abnormal

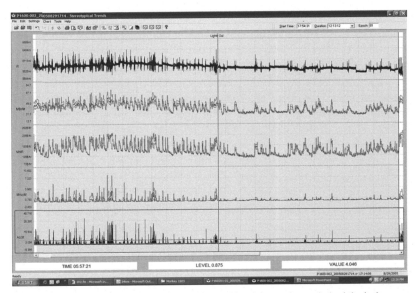

FIGURE 1    A representative vest-generated tracing of various motion and physiological parameters in a normal rhesus monkey in a vivarium environment. Note that this animal is more active on the left half of the graph, when room lights are on, versus less activity (i.e., sleep) displayed on the right half of the graph when the room is dark (nighttime).[12]

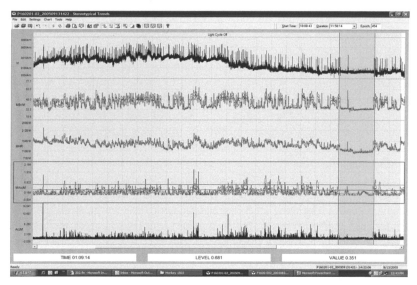

FIGURE 2   A representative vest-generated tracing of the same parameters from a rhesus monkey with stereotypy over the same length of time. Note the high level of activity on the right half of the graph (during darkness), except for a much shorter period of rest represented in the vertical bar shaded in turquoise.[12]

repetitive behavior resembling trichotillomania but probably not a stereotypy as applied to humans.[14])

On rare occasions and for reasons unknown, stereotypy can worsen into self-injurious behavior (SIB) where the animal scratches or bites itself so often and so severely that wounds appear that require veterinary treatment. I've had experience with several laboratory macaques presenting with SIB that required closure of open lacerations they had created and would not allow to heal on their own. Sometimes the animal would tear out the sutures almost as soon as they recover from anesthesia. The veterinarian is then left with a difficult choice: either to restrain or calm the animal enough so that its wounds can heal and perhaps not reoccur, or to consider the outcome hopeless and euthanize the animal to prevent further suffering and pain. Keeping the animal confined in a special chair for a few days so that it couldn't tear off its bandages was often the only physical restraint option. Drug strategies involved administering barbiturates, benzodiazepines, selective serotonin reuptake inhibitors, etc. to try take the edge off, manage the pain, reduce both mental and physical hyperactivity for a few days or even longer, and hope that when the drug wears off or the animal is returned to its cage from its chair, it will leave the old wounds alone and won't immediately create new ones. If the refractory SIB patient is a long-time research subject, as was true for one case for me, and has to be euthanized, then the loss is doubly tragic since months or years of experimentation may need to be repeated on a second monkey.

So for lots of reasons, stereotypy in monkeys needs a fresh look to reduce its incidence and stop it from worsening to SIB. Around the same time we were employing the vest to depict stereotypy in more quantitative terms such as in the two graphs above, some insight was also gained about treating it. There was ample evidence then that providing more variety to an animal's experience while in captivity could sometimes reduce or even reverse stereotypic behavior in macaques. For example, giving them puzzle feeders or moving them from indoor to outdoor housing was demonstrated to reduce the severity of repetitive motions.[15] And varying the sounds, sights, smells, tastes, and touches an animal experiences in captivity and providing it companionship with conspecifics (all falling under the practice of "environmental enrichment" [EE]) is now standard practice in laboratory animal facilities. A couple of macaques with stereotypy or SIB under our veterinary care were in behavioral studies, in which their brain responses to objects on a viewing screen were tabulated. The screen literally resembled a video game as the objects appeared in different shapes and colors and moved in various directions and velocities. Jennifer Camacho, our department's keenly observant EE manager, noticed that these monkeys appeared to undergo a reversal of their clinical signs after they were enrolled in this experiment. Such spontaneous turnarounds were theoretically possible but weren't seen in our colony or commonly reported elsewhere for macaque stereotypy. It wasn't surprising in one sense—give an intelligent animal that may be prone to clinical boredom something mentally engaging to do so that it will be less likely to direct aberrant and injurious attention on itself—but it did open our imaginations to more ambitious behavioral therapy for similarly afflicted monkeys, whether or not the underlying cause of the stereotypy was boredom or something worse (e.g., a simian form of obsessive–compulsive disorder [OCD] or autism[16]). Plus, in retrospect, the concept of employing video games for NHP mental health wasn't far-fetched. Researchers have since shown that playing friendly video games reduced symptoms and other indicators of depression or anxiety and improved mood in human patients.[17] If it works for man, then why not consider it for monkeys?[18]

Most NHP housing rooms today broadcast music and even videos for animals to hear and watch, respectively. And cages are usually equipped with plastic balls and other containers into which a few peanuts and other treats can be placed for monkeys to extract and enjoy. But are these still too passive to be adequately stimulatory, given the mental capacity of monkeys? Could more stimulation of a benign nature be even better? At MGH, I was involved in a pilot medical imaging study of macaques with stereotypy in which fludeoxyglucose positron emission tomography was performed to measure differential glucose metabolism in areas of the brain of afflicted versus normal control animals. Our preliminary findings showed increased activity in areas of the prefrontal cortex similar to that displayed in patients who exhibit comparatively similar behavioral anomalies.[19] Given our initial observations about the monkeys in the brain experiments above, we thought why not routinely provide species of laboratory monkeys thought to be susceptible to stereotypy and SIB with an even more interactive modality that includes visual, auditory, and even tactile components? If we were on the right track, this could both reduce the degree of either condition as well as possibly prevent their onset by giving the animal more to do and think about.

Practical considerations arose for converting concept to reality. Almost all equipment common to laboratory animal housing and husbandry has to be durable, affordable, safe to both staff and animals, and easy to operate in order to be acceptable for everyday use. In addition, any device accessible to macaques has to take into account their propensity to destroy any object that can't withstand their strength and persistence. We paired up with Britz & Company, an innovative cage manufacturer, to design and build special macaque housing to expand the repertoire of EE experiences and named it with a snappy acronym, the CHOICE cage (for "Cognitive Housing Options In Captive Environments"). This cage allowed the animal to control an infrared heat source if they wanted to be a little warmer. It also gave the animal control over how much socializing they wanted with a paired conspecific. If one monkey wasn't comfortable enough with the monkey in the adjacent cage, it could move a panel or wheel to close off physical contact. Conversely, it could increase access for more touching and grooming exchanges, so if and when the two animals were eventually paired within the same space, the chances of their compatibility were much greater and psychological intimidation or physical fighting was avoided. The CHOICE cage also came with a video touch screen that gave a monkey the ability to "play" with lighting elements of different shapes and colors plus an electronic drawing pad, thereby engaging hand–eye coordination and instant feedback. There was even a small angled porch at the front of the cage so the animal could get another view of the room besides just head-on.[20]

Initial trials with normal monkeys were encouraging. The CHOICE cage proved easy to operate and run through a rack washer, withstood whatever physical abuse a monkey could apply, and didn't appear likely to injure either the occupants or personnel. Interestingly, the test subjects spent the majority of time while awake close to the infrared heat source after they learned what it was, even though the room's ambient temperature was maintained within the regulated range.[21] We had grand plans to deploy these cages throughout our program and enroll monkeys with stereotypy or SIB pronto. As luck would have it (fortunately for the macaques), the number of available clinical cases shrank to zero just as we were ready to employ our prototype for this purpose. We also tried but failed to acquire stereotypic or SIB macaques from other research institutions for enrollment in an NHP behavioral therapy center of sorts. Consequently, the prototype CHOICE cage was rotated into routine EE use and not thoroughly evaluated for its presumed veterinary benefit while I was there. I still believe an apparatus such as this is worth pursuing for laboratory monkeys with stereotypy, SIB, and other behavioral problems that are refractory to other restorative approaches.

Sometimes NHP anxiety and conflict are unintended consequences of good intentions. The federal Animal Welfare Act was amended in 1985 to require research institutions registered with USDA "to provide for the psychological well-being of non-human primates."[22] When this amendment was enacted, no one was sure what it meant exactly or how to comply with confidence for animals with high intelligence maintained in close quarters. Regardless, most everyone gave it their best shot. One approach that got a lot of initial momentum was housing monkeys in groups of two or more per cage because these animals were found in social troops in the wild and it was deemed obvious that they would benefit from having a roommate. So laboratory animal care programs and IACUCs started co-housing monkeys of the same species, i.e., rhesus with rhesus, cynomolgus with cynomolgus, baboon with baboon, etc., and of the same sex so that inadvertent breeding wouldn't occur. But it wasn't appreciated enough that wild troops of monkeys are stable only after much jostling and fighting, sometimes to the death, and losers were occasionally banished from the group. At least three inconvenient circumstances in a vivarium make attaining troop stability through requisite conflict difficult if not impossible, at least for Old World species. First, few, if any, of these animals are likely to arrive with any familiarity with each other, so there's no history of prior jostling or accommodations. Second, a cage is only so big, so the jostling can't be avoided in such relatively close quarters. And losers can't leave for the jungle or savannah but are stuck in the same cage or same room as their dominant "partner," likely generating even greater distress in the absence of a lonely yet safe escape route. Third, a room of laboratory monkeys probably has higher turnover than the average wild group, so some measure of social stability may never be achieved. Just when a pair becomes compatible, one of the cage mates may complete its experiment and be euthanized to collect valuable tissues, and the adjustment process for the remaining cage mate has to start all over again.

Yet despite all of these complications and others, such as trying to make sure every monkey in a group cage gets a nutritionally complete diet every day even if it is not the dominant one, regulators persisted and programs had no choice but to comply by pairing monkeys that were previously strangers to each other, regardless of the risks. It was no surprise then that lots of fighting and trauma ensued. In one program with which I'm familiar, it initially had just as many failed attempts at pair-housing macaques as successes, even though monkeys were given weeks to get acquainted in adjacent cages with visual, auditory, and tactile access before they were placed in the same cage. Since then, safe and effective strategies for socializing laboratory monkeys have been honed.[23] These strategies should continue to be championed despite the extensive effort required, lest we forget from the old days that housing monkeys individually is usually associated with a worse outcome.[24]

On a similar note, many programs have had difficulty in pairing or group housing male rabbits, sometimes resulting in serious fight wounds; they're just not socially compatible unless they're from the same family or raised together prior to puberty. But because rabbits are observed in groups in the wild, social housing of rabbits was similarly determined to be beneficial for them and paired housing of males had to be attempted for compliance purposes, despite likely adverse outcomes. Thankfully, regulators, accrediting agencies, and IACUCs seem to have backed off since then and don't insist on paired or group housing where it's common knowledge that the animals in question are likely to fare worse. Nor is the matter of group versus individual housing applicable only to terrestrial species. There's recent evidence that even zebrafish can adversely affect one another when housed in the same tank. When maintained together for only 1–5 days, differences were detected between dominant and subordinate fish consistent with physiologic stressors. Fish that were dominant had faster rates of growth and sexual development, whereas the subordinates had higher stress hormone levels and potentially compromised immune systems.[25] It's not practical to house zebrafish singly, and there may be benefits to group housing in spite of adverse consequences of social hierarchies in a given tank. Given that the zebrafish is the second most numerous laboratory vertebrate today and continuing to grow in popularity, it would be good to know more about this when designing studies and interpreting zebrafish-derived data.

Staying on the topic of group housing a little longer, sometimes the effects of such groupings are completely unanticipated and raise more questions than they answer. Several years ago, researchers at a pediatric hospital were developing new methods of treating second-degree burns in children. They were using an established animal model in which rats were anesthetized before a heated metal surface was applied to bare skin for a prescribed time period. The animals were then bandaged and the rate of skin healing was measured histologically after the animals were euthanized. Some of the rats were group-housed, some were housed alone, and some cages were provided nesting material in addition to the usual bedding. Totally unanticipated was the finding that rats housed in groups or singly but in the presence of small cotton gauze squares provided as nesting material (known as "nestlets") healed faster than rats housed alone in cages with no nesting material.[26] Unlike laboratory mice, laboratory rats don't show much interest in nesting material, at least in conventional caging; they usually ignore it and pay more attention to hard plastic gnaw toys, whereas mice exhibit just the opposite preference. I was invited to weigh in and a second experiment was designed to see if the cotton material in the first experiment was the apparent surrogate for group housing or if alternative nesting substrate of soft wood pulp substrate would have the same effect. We found that the presence of either nesting material in the cage of individually housed rats was associated with a mildly accelerated rate of wound healing comparable to group-housed rats, versus individually housed rats with no nesting material in their cages.[27] Other authors on this paper who were experts in early childhood psychosocial development and familiar with the deleterious medical consequences of infant deprivation weren't surprised that young isolated rats healed more slowly. But the compensatory effect of either nesting substrate in the absence of cage mates was unexpected and merits further investigation.

What if environmental enrichment undermines the scientific purpose for which these animals are used? Are there cases where a less enriched environment provides a more reliable setting for consistent observations and data? The answer to the second question is yes, one I learned the hard way in an earlier job. Our departmental EE program was just getting going, and we thought it would be great to provide all mice with nestlets to supplement the hardwood chip bedding their cages contained. Within minutes after tossing a nestlet into a cage, mice start teasing apart the square to unravel the gauze, and soon they have a fluffy ball of cotton in which to sleep. In this case, everything was fine except for the cardinal sin we committed in making this change—we didn't consult our customers first. An investigator who was studying a neurodegenerative disease in mice started to get different results right after we started adding nestlets to every cage, including hers. The mouse model she had developed previously exhibited a gradual onset of clinical signs of that disease via a predictable timeline. That was critical to all her future experiments to see if new interventions had any effect when compared with an established baseline of disease progression. All of a sudden the timeline became longer and observational data were more variable. After we met with her to investigate and discovered the only change was the newly added nestlets, we agreed shamefacedly to remove them from her cages immediately, and her disease model promptly reverted to its prior dependable progression. I learned later about other published accounts of the same differential and potentially disruptive effect that EE can have on behavioral models of neurodegenerative disease.[28]

If EE interferes with the research, should it be avoided even though the animal subject likely benefits from it? Nowadays, the intentional withholding of EE usually requires explicit IACUC approval for a given protocol so animals aren't housed under Spartan conditions without scientific justification. That's not to suggest a nestlet should automatically be placed in every mouse cage. Nestlets come with their own problems, such as causing injuries by getting caught in toe nails or wrapped around limbs or tails. Their strands can get entangled in hardware on the heads of mice for some types of brain studies, and also gum up bedding dumping stations and cage wash machines. In addition, nestlets come in more than one size and, if too small, may not provide a large-enough nest to keep mice warm in a room that's cold to them. Fortunately, there are alternative nesting materials plus small disposable or reusable huts available for the same purpose. But the question remains—what scientific criteria justify a reduction or absence of EE?—and is answerable only on a case-by-case basis.

Sometimes EE can lead to positive discoveries that enhance research and our understanding of biology, such as the rat burn wound study mentioned above. Cao and colleagues reported in 2010 a remarkable and unanticipated resistance to two types of tumors implanted in mice that were housed in a veritable rodent amusement park. This enclosure was much bigger than the conventional shoebox-size

cage, holding 18–20 mice for this study and stocked with "running wheels, tunnels, igloos, huts, retreats, wood toys, a maze, and nesting material"—what I'd
call Super-EE. By contrast, mice of the same genetic background and from the
same source but housed in standard shoebox cages had much faster growth of the
same type of tumor. Even more amazing was that serum from the super-enriched
mice had just as much anti-tumor power on cancer cells in vitro and when injected
into tumor-bearing mice maintained in standard housing.[29] The authors demonstrated that the Super-EE enclosure was associated with an increase of brain-
derived neurotrophic factor in the hypothalamus that, in turn, reduced the level
of leptin in fat cells which presumably resulted in an anti-angiogenic effect on
tumors. The authors concluded that this difference was not simply a result of more
physical activity, but they didn't mention what, if any, EE was provided to the
control group of mice in standard caging. Even though positive mental rather than
physical stimulation was identified in the paper as the basis for cancer resistance
in this experiment, it's still possible that the end result was influenced by both factors. To wit, a more recent paper from a different laboratory showed a causal relationship between voluntary exercise and cancer inhibition in mice, explained by
local recruitment of natural killer cells for enhanced tumor inhibition as a result of
increased physical activity.[30] Could the causal phenomena reported in these two
studies be additive or even synergistic? In any event, it's clear that giving mice
more to do in the confines of their cages may profoundly affect research data in
addition to benefiting the animals. Furthermore and in accordance with the theme
of this chapter, offering more types of activity to which mice or other species
show a preference may be therapeutic if these animals are presenting with clinical
signs compatible with mental illness.

There's similar evidence that even the tiny zebrafish may gain from EE
added to its laboratory housing. Usually no items are placed in zebrafish tanks,
leaving more space for the fish to swim as well as making it easier to observe
them. This contrasts with the freshwater drainage ditches in south Asia, rich
with effluent and vegetation, in which zebrafish naturally live. Whether it's
intended or not, sometimes the interior surfaces of plastic zebrafish tanks
become covered with algae which make observation of the animals difficult.
But I've often wondered if that's actually better for the fish since it more
closely resembles their natural habitat. To that point, merely placing plastic
grass in zebrafish tanks increased the number of eggs produced by breeding
pairs versus tanks to which plastic leaves were added or remained barren. And
breeders that were provided plastic grass produced more fry at a younger age
than breeders that were provided plastic leaves that produced more fry when
older; regardless, neither EE had an effect on fry survivability.[31] But did these
EE objects have any stimulatory or calming effect on the adults? It would be
interesting to know if zebrafish breeding tanks with no plastic EE materials but
coated with algae along the inside walls correlate to reproductive outcomes
that differ from barren and transparent tanks? In either case, does a more naturalistic ("enriched") environment, even in a small volume for a lower order

vertebrate, have favorable biological consequences that benefit both the laboratory animal and the investigator?

All of the above discourse is based on empirical approaches to positive outcomes that target behavior modification. What if we could apply more sophisticated diagnostic tools of contemporary (human) psychiatry to identify mental illness in laboratory animals more accurately and then use the same tools for tracking responses to more sophisticated therapies of contemporary (human) psychiatry? Such therapies could involve behavioral, social, pharmacologic, and other approaches, based on the innate similarities between human and animal affective disorders of the brain. Others have recently argued to "emphasize the importance of shifting from behavioral analysis to identifying neurophysiological defects, which are likely more conserved across species and thus increase translatability," driven by continued discovery of new risk genes for psychiatric illnesses that likely apply across species (albeit viewed from the conventional perspective of animal modeling).[32]

For example, scientists have identified genetic variances that correlate with clinical anxiety and depression in rhesus monkeys and with OCD in dogs, all of which hopefully may improve our understanding of genetic influences on the corresponding human afflictions.[33] Given these promising leads, how soon will we routinely be genotyping individual laboratory animals that present with pathological behaviors for better veterinary care, rather than solely to model human psychiatric diseases? Consider having the ability to screen a colony of animals genetically to determine which ones may be predisposed to clinically damaging behaviors—a simple genomic blood test may suffice, such as markers circulating in the blood for early-onset major depression in rats, bolstered by initial success in differentiating this diagnosis from other problems in teenagers.[34] In this approach, animals at the greatest risk could then be provided a more appropriate environment to avoid or mitigate later illness, and veterinarians could devise more targeted treatment plans for pathological behaviors that arise unrelated to these animal's experiments. Today's wait-and-see-or-hope approach could seem crude by comparison. At the same time, scientists would be better served by knowing in advance if the animal subjects they've assigned to experiments are less or more likely to exhibit behavioral abnormalities later that could affect their data and conclusions.

On another technology front, powerful neuroimaging devices have provided important insights into how brains break and how they heal, thereby advancing the practice of both neurology and psychiatry. At some point in the future, these devices may be able to identify brain changes that consistently and confidently correspond to chronic depression, severe anxiety, etc. If the same brain changes can be detected in non-human animals that spontaneously develop the same or similar conditions, scanning their brains for a confirmatory diagnosis or tracking responses to therapy could enable better veterinary care. Just imagine scanning the brains of animals during a presumably distressing phase of an experiment for evidence of sustained and severe anxiety or depression. That could in turn determine more accurately if a given protocol's procedures or endpoints were truly distressful or not, if the adverse effects were temporary or long-lasting, and if and when animals should be removed from the study for humane reasons. In addition, various husbandry elements and routines could be better evaluated for potential or actual psychological benefit or harm to susceptible individual animals or the entire colony by scanning representative subjects at appropriate times. And finally, postoperative care and analgesia regimens could be judged as to whether or not they were actually effective in alleviating pain and distress. Here's hoping there's accelerated progress on this front for our aims.

(Neuroimaging has yielded other important insights that are just as intriguing. It has improved our understanding of how different persons learn differently, especially for the learning impaired. This knowledge has been helpful in discerning which approaches to teaching may be more effective for different categories of learners.[35] I'm not aware of any corresponding published research in our field about how individual NHPs or other intelligent animals may learn differently from their conspecifics, or how to identify easily those that differ. But that approach could be important for neurobehavioral protocols that require extensive training of animal subjects to perform specific tasks, making enrollment and training more efficient and effective than the usual trial-and-error method involving one animal at a time while hoping the lessons stick.)

In addition to more precise diagnostics in the future, what may psychiatry serve up as better therapies for treating laboratory animals that spontaneously develop unintended mental illnesses? There is good reason to be enthusiastic here, too. Deep brain stimulation (DBS) is a modality that has been seemingly miraculous for treating some patients with Parkinson's disease (much of which is due to insights gained from animal models)[36], and there is a lot of interest in its efficacy for depression and other psychiatric conditions.[37] Veterinary neurosurgery may sound extreme for a laboratory animal with behavioral problems. But think back to the refractory NHP SIB case mentioned earlier that had no other options besides euthanasia. If this animal still had months or years left on its protocol to generate new data and there was even a small chance of success, why not consider DBS as a veterinary procedure to restore normal mental and physical health (and sparing a replacement animal)? And not all novel psychiatric treatment modalities have to be invasive, either. It's been known for years that some chronically depressed patients feel better, albeit temporarily, after undergoing magnetic resonance imaging (MRI) scans for other medical problems. So there is a lot of interest in using MRI scans and other non-invasive neuromodulation modalities to treat drug-resistant depression and other psychiatric problems (interestingly, there's less known about these new modalities in animal models than in human patients, so enrolling spontaneously afflicted laboratory animals in veterinary clinical trials for this approach could be informative on many levels).[38]

I'm certainly not suggesting that laboratory animals with clinically severe behaviors routinely get tested genetically, be subjected to neurosurgery, or undergo brain scans (for this last approach, you may be thinking how one is supposed to get an animal to lie still in a scanner while it's being treated—perhaps it can be anesthetized for the scan and doesn't have to be conscious to derive measurable benefit). For one thing, only a few research institutions will have access to such resources for mental health assessment and treatment of laboratory animals. Neither will it be practical nor affordable to explore these options in many cases. Even more important, there are still plenty of opportunities for progress along more conventional lines of diagnosis and treatment, involving environmental enrichment, conspecific socialization, new psychotropic drugs, as well as devoting more time and gentle handling to animals that respond positively to simple tender loving care. But if the opportunity presents itself, we're obligated as laboratory animal veterinarians to consider all diagnostic and therapeutic options, for both better animal welfare and better science.

In concluding this chapter, it's exciting that we're on the verge of rightfully applying the concept of mental illness to laboratory animals that may be prone to pathological behaviors or actually exhibit such conditions. And it should not be considered a novel medical dimension, given that many others in related fields have been alluding to the comparative aspects of mental illness between humans and other animals, from Charles Darwin's time to the present day.[39] In any event, it's only fitting that new medical advances for human psychiatric diseases that were discovered and perfected using animal models be offered to animal research subjects in a purely veterinary (psychiatric!) mode to alleviate unintended suffering. To wait any longer would be a shame.

# Part III

# Laboratory Animal Management

# Chapter 10

# Is the Veterinarian Program Director a Threatened Species?

Before I started as director of the laboratory animal care program at MGH in 2002, I had already recruited my replacement. That's not to imply I had reservations about taking the job or planned to flee if things got rough, but just the opposite. One of my management tenets has been to hire my replacement as soon as possible, especially when things are going well.[1] That way, someone else in whom I entrusted the organization could make sure the trains ran on time so I could devote more attention to strategic and political issues that are just as important but often ignored in the everyday noise of running a department. Another benefit of this philosophy has been that when I eventually leave, the program will remain in good hands while a search is launched for my successor (who, conveniently, could be the person who has been overseeing the regular stuff the whole time).

One of my earlier career stints was in contract research, where commercial clients would pay our for-profit laboratory to evaluate their candidate drugs, vaccines, and medical devices in laboratory animals for efficacy and safety prior to seeking FDA approval to commence human clinical trials. When I began, we also had large contracts with the NIH to evaluate the toxicity of various environmental, consumer, and industrial chemical in rodents and to screen candidate contraceptive drugs in various species. The CRO industry is a tough business. To begin with, the underlying economics make profitability difficult—you have to maintain a modern animal facility and a skilled staff that's capable of performing complicated and lengthy studies, costing up to seven figures each, regardless of how full or empty that facility may be at any given time. Clearly, the more ongoing jobs you have to cover fixed expenses, the better, and I quickly came to appreciate that keeping the facility close to full occupancy was the only way to avoid financial calamity (think of running a luxury hotel that has similarly high fixed costs and you get the picture—every empty room is a lost source of revenue on any given night). But even if you were able to maintain a high number of concurrent studies, the net profit margins were relatively modest, only 10%–15%.[2] When I left the CRO industry after 10 years, I often wondered why anyone would go into a high fixed cost, low margin business. Wouldn't you ideally want just the opposite, i.e., low fixed costs and high margins?

Notes in the Category of C. https://doi.org/10.1016/B978-0-12-805070-5.00010-2

Another challenge of working in a CRO is that every client is equally and absolutely your most important one. This is a service business that doesn't manufacture a product. So you're creating value in real time rather than making something that can be boxed and shipped later. In addition, your reputation and likelihood of getting repeat business is only as good as your last job for that client; previously successful studies delivered on time and without mistakes don't matter if the current contract encounters delays or lapses, even when the client alters the testing protocol while it's ongoing, which occasionally happened. And in many cases, each client has staked its market valuation with its shareholders on hitting critical milestones, such as completing the pre-clinical testing of a new drug, so the client can file a New Drug Application with the FDA by the time that was promised to shareholders (and to stay in the race with its competitors). Consequently, the entire organization has to stay on its toes and you sometimes feel as if you're on a knife edge every day. It's a thrilling environment in which to work if you enjoy this sort of thing, but it can be pure hell if you don't.

For these reasons, people who have flourished in a CRO setting have always impressed me, and I felt fortunate to find Donna Jarrell when I was seeking an associate director for my upcoming stint at MGH. Donna is a board-certified laboratory animal veterinarian who was the Attending Veterinarian and program director at my previous CRO, but we missed overlapping shifts there by several years. I was told by those whose judgment I trusted that she had done good things there and that was all I needed to hear. I sought out Donna, who was working as the program director of a local biotech company, and she eventually agreed to take the plunge with me at MGH. Even before we started, we conspired to pursue two strategic objectives for the department we were preparing to run. First, we would manage the animal care program at MGH like a business, even though it was embedded within a huge non-profit teaching hospital. Both of us had years of first-hand management experience in the for-profit sector and appreciated how a "business" attitude emphasized customer service and financial discipline. The laboratory animal care program at MGH had weathered multiple leadership changes over the previous 10 years, and we felt it would benefit from a simple and consistent focus on what's important, i.e., enabling the hospital's science while staying within our budget.

Our second strategic intention was in response to how "tired" the model of a program director had become while other industries had adapted their management approaches to changes in their markets. Directing a program of laboratory animal care in any research institution historically involved administration rather than management, the former distinguished by executing a prescribed and universally established list of responsibilities while the latter encompasses a variety of moving parts that can change over time. These "parts" include animal care and veterinary staff, researchers and their scientific fields of inquiry, animal and equipment vendors, information technology platforms, facility and data security, complementary support entities such as occupational health and safety, human resources, finance, internal and external regulatory nuances, and occasionally public affairs if the head veterinarian is called on to explain or defend animal research to the media. Donna and I felt that the bountiful funding for biomedical research enjoyed in the United States for decades was not sustainable, and since animal care programs and facilities are high fixed cost enterprises, just like CROs, they would especially feel the pain if and when research funding began to tighten. If we were correct, it was reasonable to expect that in a future time of tightening finances, all of these moving parts would become further burdened with the directive from above to do more with less. Thus, we envisioned an animal care program that would be agile enough to adapt to anticipated fiscal constraints while not compromising animal welfare, worker safety, and investigator needs. If we were wrong and the hospital continued to enjoy robust research revenues while remaining fiscally strong in an era of capped health care reimbursements, then at least we would have the satisfaction of eliminating wasteful expenditures while we resumed running in place on the treadmill of convention. Plus, this would be a lot more fun irrespective of how it turned out.

After a couple of years to getting acquainted with the animal care staff and the "MGH way," we started shopping for attractive operations management alternatives. Leaning on the lessons we learned from working in the cauldron of a CRO, we knew that adaptability would be a key element of success but didn't know where exactly to turn. During my CRO years as the general manager of an entire operating unit, I had enrolled in a 10-week international executive education program at the Harvard Business School (HBS) that provided valuable insights into management practices of many industries around the world. So we sent Donna to HBS in 2004 for a 1-week session devoted to operational excellence, followed by an 8-week program on general management in 2006, so she could get a sense of what was out there for our needs and her personal development. She came back inspired by lean management and continuous improvement as practiced by the Toyota Motors Corporation and embraced in many other industries as the Toyota Production System (TPS).[3] A primary goal of TPS is to eliminate any activity or cost that does not create value for the customer. In our case, the customers were scientists who paid us from their research grants for husbandry and veterinary care for their animals. Everything about TPS that Donna shared made immediate sense and so we were off. Because TPS principles are described in quite simple phrases, we made a common mistake of equating simplicity with ease of adoption, and our staff had to endure multiple false starts and dead ends during the next couple of years.

Donna then learned about a consulting firm that specialized in lean transformations, and we retained their services for a six-figure contract to embed themselves in our operations for 14 months. We were confident that the investment in money and time, while not cheap, would pay for itself in several years or less based on how much waste we felt was entrenched in our program.

Around the same time this was all happening, NIH research budgets were getting squeezed, even before the Great Recession of 2008–09. Compared year to year and in contemporary rather than adjusted dollars, the NIH budget funding for research grants was already in the decline, dropping from a 10% increase between FY2002 and FY2003 to a 1% decrease between FY2005 and FY2006.[4] Arlen Specter, the late US Senator from Pennsylvania, came to the rescue by championing an additional allocation of $10 billion for NIH-directed research, research training, and new extramural research infrastructure in the American Recovery and Reinvestment Act (ARRA) of 2009.[5] But the rapid deceleration of NIH funding over the years immediately preceding Senator Specter's heroics had reduced everyone's bullishness about money for biomedical research. Even gloomier, the ARRA respite was only temporary, and no one knew what would happen later if the United States was still in the depths of the recession. Thus the word went out that this funding environment was the new normal, equally applying to federal, corporate, and foundation research grants, and to get used to it. Donna and I didn't consider ourselves clairvoyant or lucky by guessing correctly. But we were relieved that we had begun establishing a department culture of thrift and efficiency when we did. Initiating and sustaining change in any organization takes years and we didn't begin this transition a moment too soon.[6]

Why am I recounting these stories here? Because they're presented as examples of how important program leadership can be under the inevitability of change. One of the characteristics of discovery science is that it never stands still. Why should we comfortably presume its evolving needs will not similarly demand changes in how laboratory animals are cared for? Even if no significant changes are forthcoming from investigators, other new knowledge about animal biology and welfare that will surely emerge are just as likely to require changes in husbandry and veterinary medicine, albeit possibly of a different nature. By the same token, why should we presume that yesterday's program leadership model of (passive) administration rather (proactive) management will suffice as research and research animal models evolve?

This perspective may sound insultingly obvious until one examines how program directors are traditionally prepared to direct programs. It's almost always an apprenticeship, relying on osmosis rather than explicit instruction about managing, at least beyond the rudiments of regulatory compliance and building an annual operating budget. How about the management training provided by post-doctoral training in laboratory animal medicine, a popular entry point for veterinarians who want to specialize in the field? The emphasis of these training programs is usually clinical experience or individual research projects, and rarely includes management details beyond exposure to the inner workings of an IACUC and preparing for an AAALAC site visit (again, stressing compliance as if that was the only or most significant element of management). Some training programs have begun to include an optional or abbreviated internal seminar series on program management for post-docs, but executive managers from the outside are rarely brought in to speak. That's likely because program directors don't get much, if any, general management training themselves. So why should they know which subject matter to emphasize or how to present it when general management is the topic? Consequently and unless things change, the next generation of program leaders will remain just as ignorant about how the many other dimensions of management can influence program quality, efficiency, finance, employee development and retention, customer satisfaction, and the ability to adjust to changing externalities. Directing a laboratory animal care program involving a large staff and millions of dollars in annual operating expenses may be the only job besides running for political office that requires no formal training or demonstrated competence in general management even though both of these occupations benefit from management know–how.

How could I reach such an insolent conclusion? After all, I owe my professional credentials and livelihood to the benefits of this same training legacy. Aren't programs doing just fine, or at least well enough, without placing yet another burden on a director's already overflowing plate? Why am I convinced this is so important? For starters, it's been my experience from programs I've directed as well as those I've evaluated for others that the average program of laboratory animal care wastes around one-third of its time and expenditures on activities and items that don't create value for the customer and, hence, aren't necessary. Many programs are now being pressured by their host institutions to cut expenses without jeopardizing any other components. How much of a difference could enlightened frugality make?[7] If laboratory animal care programs were managed more efficiently to eliminate the percentage of their budgets that is wasted, imagine the impact that this additional money could potentially have on biomedical discoveries. Which medical riddle could be solved or which new scientist would be trained as a result of the newly available funding?

But unless you already have management training and experience in reducing costs, how could you know where to start and still succeed on all the other fronts while the clock is ticking? And if the potential financial upside is so sizable, why haven't programs been forced to embrace thrift long ago? Three common avoidance tactics proved useful in the past. The first plays the compliance or accreditation card, based on fear rather than logic. It goes something like this: no matter how pricey the program may be, sustaining its expenses is the only way the institution can remain on the right side of the rules or of AAALAC.[8] One would shrug sympathetically when explaining this to the institution's accountants while blaming outsiders for insisting on ever more expensive regulations with no options. This approach worked as long as other programs at other institutions didn't break rank and demonstrate that regulatory compliance or accreditation and cost cutting are not mutually exclusive. The second tactic involves timing non-capital purchases around the end or beginning of the fiscal year. If money is still "left over" as the current fiscal year is ending, i.e., if it looks like we'll finish the year ahead of budget, let's spend it on items we may or may not need before the new fiscal year ushers in a new budget so that at least we have them around in case we ever need them (this also serves to protect my annual budget from cuts because I spent ("needed") the funds). The third maneuver resembles the old shell game, shuffling expenses from a category or fund that is maxed out to one which still has some room for hiding more purchases, provided the definition or purpose of the other category or fund is fungible. Certainly, there can be credible reasons driving any of the above accounting transactions. But as financial screws tighten and program spending is scrutinized more intensively, these transactions will have to be truly legitimate if the director wants to remain credible and employed.

This leads us to the second reason why all of this is important. Until recently, veterinarians were the only ones considered qualified to be program directors. Others who weren't veterinarians could rise only to the title of Assistant Director or Associate Director solely because they weren't veterinarians. These persons oversaw most of the administrative duties of a program, usually because they had more administrative training and were better at it than the veterinarian director. But as we transition from passive administration to proactive management to keep a program vibrant and effective on multiple fronts, will we witness a reconsideration of what are the most important attributes a program director should have? As executive management skills become more critical, as I believe they will, for successful program leadership, why would a candidate with little or no executive management background even be considered for the position? If I'm doing the hiring, why shouldn't I seek only a seasoned senior manager who has a proven track record of success in such disciplines as strategic planning, finance, organizational development, and even marketing? He or she can then hire a veterinarian to do the usual veterinary things in a diminished (and more affordable) role. A few programs around the country already have non-veterinarian directors. From the outside, they seem to be doing no worse if not better than programs with veterinarian directors on the basis of regulatory compliance and AAALAC accreditation, the two old indicators of acceptability.

Veterinarians can still retain their status as best qualified to direct programs of laboratory animal care. They already have the scientific and technical expertise requisite for providing high-quality animal care and understanding animal pain and distress. Those with the title, Attending Veterinarians are the only persons at their institutions, at least in the United States, who can intervene without interference on behalf of any laboratory animal at any time for humane reasons. These weighty professional and ethical responsibilities complement a leadership role to oversee all aspects of laboratory animal care and medicine in a research institution, especially the large and complex ones. So how should veterinarians acquire the management skills needed to remain in prime consideration for those leadership positions?

If the best teacher is experience, then I was fortunate to get started in my CRO stretch in the mid-1980s. The only reason I took a job as Attending Veterinarian at this company was because it was the only position available in laboratory animal medicine in the region and my wife and I wanted to remain in the Boston area after my post-doctoral training was completed at the Massachusetts Institute of Technology. As is the case in many situations in life, timing is everything. Around this period, the first wave of therapeutic recombinant human proteins were being developed by new biotech companies, and they were all looking for CROs to evaluate drug safety and efficacy in laboratory animals before FDA would allow human clinical trials to commence. Since the physiological properties of these proteins were already known, the questions for pre-clinical testing protocols centered around how would these recombinant proteins behave at pharmacological doses? Would higher-than-normal levels in the body trigger bad reactions unrelated to their normal functions? And since these were human proteins administered to non-humans, would these proteins be recognized as foreign antigens and trigger an immune response? If so, would that immune response be inconsequential or cause serious lesions? If the latter, how could one distinguish between an artificial outcome, based on the drug's circumstantial antigenicity in an animal that may not be manifested in a human patient, versus toxicity solely due to the drug itself? Addressing these and other unknowns required much more deliberation and planning for each protein than the conventional and highly uniform process for testing non-biologic drugs in vogue at that time. Each contract bid was an intellectual challenge that we tackled with gusto to win. That quickly gained us a reputation as a worthy partner in the biotech sector, and generated good business for the laboratory. Between 1985 and 1993, annual revenues from for-profit clients grew from $200,000 to $14 million in a highly competitive industry that was purported to have no room for new entrants, and total annual revenues tripled from $5 to $15 million even as our animal testing contracts with NIH receded over time.

During this exciting period, I kept moving up the management ladder simply because those above me kept leaving for various reasons. I had no master career plan or ambition but discovered how much I enjoyed the for-profit sector and client service. The best management training I got was a consequence of having "profit and loss" (P&L) accountability, with an emphasis on the "P." All the facets of general management were constantly in play, from strategy to finance to marketing and sales to supply chains and operations throughput to hiring and retaining talented staff. In addition, this CRO was a part of a publicly traded company; so our financial results were always included in the quarterly financial reports to shareholders. This provided a frequent reminder (as if I needed one) of what's important. I was promoted to president and general manager of our laboratory in 1989, after it was sold to another corporation and the prior head of the CRO, my former boss, left. In 1993, the company's Board of Directors replaced some of the senior executives with those of us managing the business units in the midst of a downward spiral. I was appointed a corporate vice president, and we were eventually acquired by a new company (our third owner in 10 years) in the course of successfully turning the business around. Many mistakes were made, including plenty of my own doing. The lessons learned were invaluable and more firmly grasped than if I had merely read books and attended seminars about management. Thus, my management skills were developed by necessity and survival and have served me well since. But I don't recommend others seek the same level of stimulation (and exhaustion) just to gain some business acumen. So where can current and potential veterinarian directors obtain appropriate management training and experience either to safeguard their jobs or prepare to be the best applicants for an opening?

Don't look to the laboratory animal care industry. That's because almost all programs have been sheltered, until very recently, from market forces that sharpen one's perspective and provide quick and prompt feedback about which management decisions work and which ones do not. This sheltering is comprised of at least three realities. To begin with, our programs don't have to compete for business, thereby eliminating a basic and effective stimulus for managerial excellence. If the program is centralized so that all researchers at that institution must keep their animals under the program, then these researchers have no other options for animal care services. Without competition, there's no incentive for better management in order to attract and retain customers. Second, almost all programs are subsidized, if not completely financed, by the institution.[9] This provides a safety net, both monetarily and psychologically, and removes a powerful impetus for tight financial management. Third and just as important, research administrators at the top are usually just as lacking as program directors when it comes to sharpened business experience. So they're interested only in macro-outcomes (less grumbling from scientists and satisfactory regulatory audits) rather than operational micro-details, and can't provide helpful guidance while at the same time insisting program directors do more with less. This is not to suggest institutional officials are nonchalant or clueless about animal care programs, only that they only know what they know and rely on program directors for smooth sailing. That's all the more justification for keen management expertise at the program level.[10] Without the incentives to compete for business while watching every cent, it's no surprise that program management as subject matter historically focused mostly on regulatory compliance and staff supervision (e.g., how to handle "difficult" employees) as important management topics.

If the laboratory animal care sector isn't helpful for this purpose, where else can one turn? A Masters of Business Administration (MBA) degree would more than suffice but involves expensive tuition and even more expensive time (at least 1 year and usually more). I don't believe a commitment of that magnitude is necessary nor would the payback on the investment be favorable because an MBA degree isn't valued enough in our field to command a commensurate salary bump. Instead, I suggest reading lots of articles in the general business press, especially about small companies since they more closely resemble the programs we direct versus the large multi-national corporations that get the most ink, and thinking how business scenarios and outcomes in these articles could be applied in one's program. At this point, I can almost hear the reader thinking "but my program doesn't assemble trucks, run a chain of hair salons, supply donor blood to hospitals, or produce light switches—we just provide husbandry and veterinary support to laboratory animals." One of the best parts about studying management is that more, not less, of the details are transferable across industries, and lessons learned are readily applicable if you think about it enough. My definition of management is simple and as follows: *matching the needs of the organization with the needs of the employee.*

This definition recognizes that each party has strong self-interests and each party usually has more than one option; management doesn't have to be any more complicated than that. Every component of an organization involves persons doing something on its behalf, so the emphasis on the employee is appropriate. Known as "soft skills," interacting with the organization's workforce is often under-appreciated, under-emphasized, and usually dominates a manager's time whether you're making a product or providing a service. Issues such as recruiting and hiring, training, levels of organization, performance appraisals, promotions and career development for the ones you hope to keep, and corrective actions for the ones who need it are fairly consistent across industries. That's simply because workers are human, with their own ambitions, needs, constraints, distractions, and abilities, regardless of their place or type of employment. How to manage all these variables, usually through subordinates, in concert rather than conflict with other departments such as Human Resources, Physical Plant, Purchasing, Payroll, and Environmental Health and Safety, relies more on such soft skills rather than the "hard" factual knowledge accumulated in biology and veterinary medicine. So absorbing teachings from other managers in other settings can be illuminating and often transferrable into situations encountered in program leadership.

A basic college-level accounting course is also recommended—even a 2-day seminar on finance for the non-financial executive is a good start.[11] One needs to learn the language of how money is classified and tracked, even though one would not be expected to balance accounting ledgers. Almost all program directors are answerable to annual operating budgets or operating income (P&L) statements. Directors should track every number in every line item and know every definition of every revenue or expense category in order to manage their programs' finances effectively. By contrast, few, if any, program directors will ever be responsible for balance sheets and cash flow statements so these dimensions are understandably less emphasized. Nevertheless, the information contained in all of these three basic accounting documents is an excellent barometer of the financial health of the parent institution and in turn influences what is expected of a program director. Knowledge of Generally Accepted Accounting Practices (literally known as "GAAP") can also be helpful if more than one legitimate option is possible in assigning an expense to different categories and if the outcome determines whether or not a necessary expenditure gets approved. In addition, being fluent in GAAP-speak and other accounting regulations and standards comes in handy when one's program or institution is undergoing a financial compliance audit. Familiarity with cost accounting also is sometimes useful when it comes to calculating per diem expenses for routine husbandry services. One standard per diem costing exercise starts with gathering time-and-motion data for each husbandry or veterinary task, a bottom–up approach of sorts.[12] This method may suffice as long as the tasks don't change during the fiscal year for which the data are applied and as long as ancillary and support-ive costs aren't ignored. But if the program is engaged in continuous improve-ment so that tasks may be revised at any time to enhance quality or efficiency, then the original time-and-motion costing data become invalid. This in turn may jeopardize the budget and result in per diem prices that are different from what researchers are willing to pay or the institution is willing to underwrite. In these situations, a top–down per diem costing model can provide more flexibility while still satisfying outside auditors. Business accounting knowledge makes financial management options such as these easier to use.

Additional methods for instilling and upgrading one's management aptitude should be explored and could be fun. Consider executive education courses given by business schools such as the ones Donna Jarrell and I attended. They don't have to be 8–10 weeks long; several days to a week can provide a deep-enough dive into a particular management discipline. I also recommend attending the most prestigious business school you or your employer can afford. That's not to impress others but because these schools are more likely to attract more interest-ing classmates working in a wider variety of organizations and industries from around the world. At this stage of one's career, interactions with other students are often just as enriching as knowledge gleaned from the faculty. And consider enrolling in management subjects that you think aren't germane to overseeing laboratory animal care. Topics such as marketing (researching your customers' needs) and sales (fulfilling those needs) may not seem relevant but could expose you to different ways of thinking about your program and create new opportuni-ties for improvement as well as personal growth.

On the subject of researching customer needs, while I was at MGH we decided to launch a customer dissatisfaction survey. I wasn't interested so much in what animal users already liked about what we were doing because that was suppos-edly what they were paying for. Instead, I wanted to know what they thought we should be doing better for their per diem payments and were there other services we didn't offer that could further support their research if only we knew about them? I also wasn't worried that a mob of angry scientists would use this survey to call for my head. While there were occasional frustrations or heated outbursts about something we did or a mistake we made, I was confident that this survey would be appreciated as an invitation to point out unfinished business. I also knew that getting feedback like this would be a good exercise for our department, both to highlight areas for renewed focus and to let everyone know the sky wouldn't fall if we asked others to describe our perceived faults. Plus the results could indirectly indicate the things we were doing that were pleasing to our customers.

Thinking about what to include in the survey wasn't difficult because unso-licited feedback on a short list of subjects was usually given when things didn't go smoothly. But this would be the first time we could obtain a representative snapshot involving more than one opinion on defined topics. To make sure we got a sufficient number of responses, we offered a $50 per diem credit for every completed survey within the time allotted. We capped how many respondents (10) could be from the same principal investigator's laboratory so that nobody could game the system. The survey results were to be ranked, starting with the most common categories of dissatisfaction at the top, and then shared with the entire animal user community for discussion and follow-up. The survey was offered via a web link with the following questions:

1. Indicate in which vivarium/vivaria you conduct your research? (check all that apply)
2. Which species do you use in your research? (check all that apply)
3. Please rate the frequency and quality of CARE (HUSBANDRY) of your animals based on the following statement below.
   I am DISSATISFIED[13] with the level and quality of:
   - Cage/Pen Conditions (Food, Water, Bedding)
   - Husbandry/Care (Scheduling, Frequency, Animal Receipt)
   - Cage/Pen Special Husbandry (Special Feeds, Treated Water, Customized Bedding)
   - Cage/Pen Management (Flooded Cage Management, Overcrowding, Deceased)
   - Veterinary Care (Treatment of Health Concerns, Notification of Health Concerns, Discovery of Health Concerns, Outbreak Management)
   - Research Support Services (Breeding, Anesthesia and Surgery Support, Rederivation)
   - Please provide more details for all "Often" and "Always" responses.
4. Please rate the frequency and quality of the FACILITY and/or the EQUIPMENT based on the following statement below.
   I am DISSATISFIED with the level and quality of:
   - Animal Room Equipment (Ventilated Cage System, Reliability of Cage Changing Hoods, Static Cage System, Automatic Watering System)
   - Animal Room Maintenance (Availability of Supplies within Animal Room(s), Availability of Cage Changing Hoods, Cleanliness of Animal Room, Caging/Water Bottles)
   - Facility Maintenance (Cleanliness of Facility, PPE Availability, PPE Standards, Hallways, Work Areas, Procedure Room)
   - Please provide more details in the box below for all "Often" and "Always" responses.
5. Please rate your experiences with our STAFF (Facility Managers, Program Managers, Veterinarians, Team Leaders, Research Animal Specialists).
   I am DISSATISFIED with the level of:
   - Professionalism
   - Knowledge
   - Courtesy

- Helpfulness
- Responsiveness
- Availability
- Please provide more details in the box below for all "Often" and "Always" responses.

6. Please rate the frequency and quality of the ADMINISTRATIVE FUNCTIONS and/or PROCESSES based on the following statement: I am DISSATISFIED with the following:
   - Animal Procurement (Online Ordering, Paper Ordering)
   - Cost (Per Diem Rates, Services/Non-Per Diem Rates, Animal Transportation)
   - Researcher Orientation/Facility Access (Animal Use Orientation, Facility Tours, Animal Facility Access Process, Occupational Health Clearance)
   - Rodent Import/Export/Quarantine (Shipping, Scheduling, Quarantine testing)
   - Animal Transport (Scheduling, Ease of use)
   - Please provide more details in the box below for all "Often" and "Always" responses.

7. Please provide your contact information below to assure receipt of any possible per diem credit for completing this survey. After successful completion of this survey, a $50 credit per survey (up to 10 per principal investigator) will be issued.

We received 184 completed surveys, representing between 5% and 10% of the total animal user pool, that nicely matched our species mix (i.e., mouse users comprised 88%, rat users 8%, and NHP and livestock users 4%). The financial hit of $9200 for per diem credits was trivial because it was barely a rounding error in our annual operating budget. Surveys were analyzed to see which questions generated the highest sum of "Occasionally," "Often," and "Always" (i.e., aggregated "Dissatisfied") responses. The results of our first survey are presented below (we repeated the exercise the following year to see which, if any, categories improved or worsened after changes were made).[14]

| Survey Item | % Dissatisfied |
|---|---|
| Cost (per diem rates, services/non–per diem rates, transportation) | 41 |
| Cage/pen conditions (food, water, bedding) | 41 |
| Cage/pen management (flooded cage management, overcrowding, deceased) | 37 |
| Animal room equipment (ventilated cage system, reliability of cage changing hoods, static cage system, automatic watering system) | 30 |
| Animal room maintenance (availability of supplies within animal rooms), availability of cage changing hoods, cleanliness of animal room, caging/water bottles) | 30 |
| Animal procurement (online ordering, paper ordering) | 28 |
| Husbandry/care (scheduling, frequency, animal receipt) | 24 |
| Facility maintenance (cleanliness of facility, PPE availability, PPE standards, hallways, work areas, procedure room) | 20 |

These results were distributed to the MGH animal user community with a reminder that since someone much higher than I in the research hierarchy set the per diem rates, it wasn't an actionable item by my department. Thus, it was struck from our list of follow-ups. Otherwise, everything else was fair game, and I engaged our program's leadership team to consider improvements in the other survey responses. Sometimes, the follow-up simply involved re-educating animal users about how equipment was to be used correctly and the shelf life of various supplies, or reminding them that we couldn't execute their orders in a timely fashion if the requested information was missing on the form. In other cases, we surveyed the users again but for specific issues such as how often they thought a rodent cage needed to be changed, even though almost all cages brought to our attention were not excessively soiled by conventional standards. That didn't matter half as much as giving investigators a voice in how their animals were maintained. The best outcome of the survey was the goodwill it generated amongst our customer base. Having the courage to ask them to point out our warts (and even paying them to do so!) left such a positive impression about how seriously we valued their opinions that I still have researchers recalling that survey years later. The lesson here is that there's value in not only aiming to please your customers but also asking what specifically doesn't please them. While surveys such as this one are perhaps a backhanded way of learning which things you're doing are going well, it's sometimes more fruitful than more conventional tactics. That's especially true if the next step is to invite the customer to help design and evaluate improvements so that they then share in ownership of those upgrades.[15] And that's precisely what we did in addressing the most common complaint about rodent cages judged by users to be "too dirty."[16]

What other options are conceivable for providing current and future program leaders more management principles and tools? One possibility that Donna and I have discussed on occasion is to convene a "Director's Camp." Our vision is to present management subject matter that we feel those in charge of programs should know, regardless of their veterinarian credentials or specialty training. One approach could be organized around conventional topics as outlined below:

*Day 1*
- The Role of Management: establish what executive management entails and why it matters in laboratory animal care now
- Effective Program Leadership: examples of how executive management can make a difference in the quality and productivity of laboratory animal care
- Strategic Planning/Setting a Vision: framing the purpose and direction of your program
- Finance: the value of financial literacy, establishing and tracking operating budgets, calculating returns on capital investments, soliciting and evaluating quotes from vendors, open book management, activity-based costing

*Day 2*

- Information Technology: the application of new IT tools in program activities
- Operations Management: how program processes can be mapped and revised to eliminate unnecessary work while simultaneously improving program quality
- From Bedside to Cageside: human medical Point-of-Care innovations for better laboratory animal medicine
- Tour of local vivaria engaged in continuous improvement with time to meet with their animal care staff

*Day 3*

- Organizational Development: the value of soft skills in attracting and retaining a strong middle management layer and respecting every employee in the program
- Workforce/Pipeline Development: how to create a "bench" of trained and ready talent at all levels of the organization for smooth transitions when staff leave
- The Voice of the Customer: testimony from animal users about what they expect from programs
- Supply Chain Management: building strategic partnerships with vendors for continuous improvement in supplies and equipment

An alternative or advanced curriculum could accentuate how dynamic program management needs to be in step with the dynamism of biomedical research. Thus a simpler and more pointed approach could be organized as follows:

- Embracing Change: how various organizations (companies) in other industries have succeeded by adapting to changes in their competitive environments, and what lessons can be learned for laboratory animal care
- Implementing Continuous Improvement: overcoming your staff's fears about change
- Measuring Continuous Improvement: how outcomes of alternative process pilot project can be quantified for easy and objective assessment by all
- Monetizing Continuous Improvement: calculating the financial savings realized through process change (e.g., labor efficiencies, reduced rework, fewer consumed supplies) to gain the support of senior research management or administrators
- Promoting Continuous Improvement to program employees (and their unions, if applicable), animal users, the IACUC, and others
- Sustaining Continuous Improvement: how to maintain momentum after initial successes and inevitable setbacks

While various conferences and gatherings include a few sessions on program management, nothing as focused and extensive such as this exists today. It would be interesting to see what interest is out there and how our peers would respond.

The purpose of this chapter has been two-fold: to highlight how complex program leadership is becoming, and to contrast those complexities with how executive management skills continue to be undervalued by laboratory animal veterinarians and their employers. We have enjoyed an implicit monopoly to lead these programs since their inception over 50 years ago. But as the financial and operational stakes get higher, more management expertise will be needed to stay competitive. Contrary to conventional wisdom, there's no law or regulation that states the head of a program of laboratory animal care must be a veterinarian, only that the Attending Veterinarian is responsible for the veterinary care program, including all elements involving animal welfare and well-being.[17] As stated above, these obligations can be just as logically subordinated under a non-veterinarian Director who has broader duties for all aspects of a program besides just the veterinary ones. Unless veterinarians seek to upgrade and broaden their executive management competencies as leaders of programs of laboratory animal care, they'll be vulnerable to displacement by professional executives.

# Chapter 11

# Sometimes "Lean" Means "More"

*When you pay attention to boredom it gets unbelievably interesting*

Jon Kabat-Zinn

Biomedical research is undergoing unprecedented financial pressures. Federal funding is flat, at best, while state government contributions to public universities and their professional medical schools are mostly down. Private foundations, such as the Wellcome Trust, Howard Hughes Medical Institute, and Gates Foundation are filling some but not all of the resultant gap. On the for-profit side, drug, vaccine, and medical device manufacturers are downsizing and outsourcing R&D while overseas CROs are taking more business from their US counterparts based on price. Programs of laboratory animal care aren't immune from these pressures and being asked or forced to reduce their operating and capital budgets accordingly. Most programs are struggling to cope because they've never had to reduce expenses or get more out of less to the extent required today. What's a program director to do?

In dealing with this new reality, I had an advantage over most others due to the 10 years (1985–95) I spent working in the pre-clinical CRO sector. This is an industry sector in which cash is rarely abundant, net profit margins are modest even in a strong quarter, and for most clients you're only as good as your last study when it comes to competing for the next one (with price always a major decider for the winning bid). In such an environment, thrift is not only a virtue but also a necessity for the survival of a CRO over the long term. Embracing thrift also came in handy when I was hired on two occasions to head up biotech startups. In both cases, for 2–3 year stints each, we established laboratories and business infrastructure from scratch while maintaining a monthly "burn rate" of no more than $200,000 during my stay at each company. While that may sound like a lot of money, it's really not much when you consider how fast you need to ramp up both the company's science and visibility in order to attract new money to supplant earlier financing rounds, hoping all the while your research moves forward while both private equity investors and multi-national corporations in the markets you're targeting remain enamored of your core technology.

Getting back to program management and as mentioned in Chapter 10, we were attracted to something called "lean" management, defined as eliminating any and all costs not of value to the customer.[1] The Toyota Motor Company is probably

Notes in the Category of C. https://doi.org/10.1016/B978-0-12-805070-5.00011-4

**105**

the most popular archetype of lean manufacturing, going from a small exporter of economy vehicles in the 1960s to the largest car company in the world today. Its approach to manufacturing has been studied and copied by other businesses for decades, resulting in what's become known as Toyota Production Systems (TPS) for other firms to follow.[2] But Toyota's success isn't only about cutting out waste. After gaining a toehold in the United States, it set out to create cars and trucks that were not only attractively priced but also more reliable than those of its domestic competitors. In addition, consider that Toyota reduced the time to bring a new model from concept to market from the usual 4 years to only two. That in turn created a huge advantage in the hybrid vehicle sector for one. Even though Toyota and General Motors announced their intentions around the same time, Toyota's Prius arrived on the market 10 years before Chevrolet's Volt, undergoing multiple model improvements during that interval. The result to date is 9 million Prius sold in the United States versus only 100,000 Chevy Volts.[3] Overall, Toyota's profits in 2015 eclipsed the comparable total for the three US car makers combined. More to the point, Toyota's earnings per vehicle that year were three times higher than for Ford or Chrysler and four times higher than that of General Motors.[4] Toyota's triumph is a testament to organizational dexterity and speed as much as it is to eliminating waste, all of which is enabled by an underlying dedication to continuous improvement as we shall explore further below.

Another example of lean management success but in a service company rather than a manufacturing one is Southwest Airlines. There had been earlier airlines offering cheap fares based on lower costs, but Southwest has been the most successful by far. Consider that since the terrorist attacks of 9/11, when airline travel plummeted, and throughout the Great Recession beginning in 2009 (when airline travel plummeted again and for a longer period of time), Southwest is the only US airline to have been profitable every year. And contrary to the conventional wisdom about the uncompetitive cost burden of labor unions, Southwest was the most unionized of all US airlines over that same period. Its lean management practices were based on passenger needs, such as quick turnaround times at the gate, frequent flights between cities within 500 miles of each other, and using less congested airports. At the same time, it eliminated activities and expenses of little or no concern to most of its customers, so only one type of airplane, the Boeing 737, was used in order to minimize parts inventories and simplify maintenance. Common in-flight amenities were eliminated, such as expensive meals and reserved, multi-class seating. The combination of all these tactics was so successful in achieving profitability while undercutting the fares of competitors that a phenomenon known as the "Southwest effect" was coined: whenever it was announced that Southwest would link another airport to its network, the other airlines serving that airport quickly dropped their fares or left because they couldn't compete.[5]

How could we leverage lean management principles for laboratory animal care? More important, how could we get buy-in from the department to change its culture from business-as-usual to lean when there was no business crisis, no obvious "burning platform," to serve as a catalyst for eliminating waste?

Despite 5 years of trying it on our own, we were still only scratching the surface of doing things better, faster, cheaper, and safer. Donna and I concluded we needed some professional guidance to really make a difference. That led us to retain a lean management consulting firm, Murli and Associates,[6] of whom Donna had learned at a lean operations seminar in Atlanta. Murli's folks jumped right in and began instructing our entire management team on value stream maps, kaizen events, Gemba walks, and visual controls such as 5S organization of the workplace and Kanban cards for managing supplies inventory. Some of the terms and concepts we already knew, but we had never so intensively focused on their meaning and implementation. Early on, several things became evident. First, "simple" doesn't automatically mean "easy." For something that sounds so obvious like 5S-ing the workplace, changing attitudes and practices was so much harder than just understanding the concept and following its instructions (see below). Second, a total commitment to lean and its consequences is necessary not only to make real progress but also to reorient an organization's way of thinking about the work it does so that any gains we made would be sustained. Finally, if program leadership isn't passionate about embracing lean as a permanent change in the program's DNA, then the whole thing's a waste of time. Personnel at all levels will resume old habits unless constant vigilance and gentle yet firm reminders are employed. Like anything else, managing is all about the people and resistance to change is a fundamental human behavior.

I'll use 5S as an example. It's called that because it involves five ascending levels of workplace organization, and the name of every level starts with the letter "S":

1. *Sort* (sometimes called Simplify): Go through items and keep only what is needed as part of standard work and remove everything else.
2. *Straighten*: Organize work spaces so everything has its designated place and all items necessary to do the work are available.
3. *Shine* (sometimes called Scrub): Always keep work spaces clean; the cleaning process also acts as a form of inspection to expose potential problems that could hurt quality or efficiency.
4. *Standardize* (sometimes called Stabilize): Develop systems and procedures to maintain and monitor the first three S's across the entire organization.
5. *Sustain*: Maintain systems that have been put in place while pursuing further improvement in each of the first four S's, i.e., a never-ending process.

These names are adopted from the corresponding Japanese names for the same objectives, each conveniently also starting with an "S" sound. Our goal was to apply 5S eventually to every room we used for animal housing, veterinary care, cage processing, and storage, as well as corridors, entryways, and shared administrative and meeting places, for a total of 270 spaces; personal offices and work cubicles were exempted, as were animal procedure space

## 5S Scoring

| | Sort | Straighten | Shine | Standardize | Sustain |
|---|---|---|---|---|---|
| Level 5 focus on prevention (score 10) | Employees are continually seeking improvement opportunities | A dependable method has been developed to provide continual evaluation, and a process is in place to implement improvement | Area employees have devised a dependable, documented method of preventive cleaning and maintenance | Everyone is continually seeking the elimination of waste with changes documented and information shared | There is a general appearance of a confident understanding of and adherence to the 5S principles. A culture of cleanliness and orderly maintenance of the workplace is expected by everyone |
| Level 4 focus on consistency (score 8–9) | A dependable, documented method has been established to keep the work area free of unnecessary items | A dependable, documented method has been established to recognize in a visual sweep if items are out of place or exceed quantity limits | 5S agreements are understood and practiced continually | Standard work is consistently followed on all shifts | The workforce is actively engaged in driving continuous improvement in 5S scores |
| Level 3 make it visual (score 6–7) | Unnecessary items have been removed from the workplace | Designated locations are marked to make organization more visible | Work/break areas and machinery are cleaned on a daily basis. Visual controls have been established and marked | Visual control and standard work is in place and proven out | Weekly 5S reviews are conducted reliably by the plant manager and others. Feedback is being acted on |
| Level 2 focus on basics (score 4–5) | Necessary and unnecessary items are separated | A designated location has been established for items | Work/break areas are cleaned on a regular scheduled basis. Key items to check have been identified | Methods are being improved but changes haven't been documented | A recognizable effort has been made to improve the condition of the workplace |
| Level 1 just beginning (score 2–3) | Needed and not needed items are mixed throughout the workplace | Items are randomly located throughout the workplace | Workplace areas are dirty, disorganized, and key items are not marked or identified | Workplace methods are not consistently followed and are undocumented | Workplace checks are randomly performed and there is no visual measurement of 5S performance |
| Level 0 back sliding (score 0–1) | Event was carried out but improvements have since deteriorated | Event was carried out but improvements have since deteriorated | Event was carried out but improvements have since deteriorated | Event was carried out but improvements have | Event was carried out but improvements have since deteriorated |

in our facilities that was for researchers. Murli's team gave us the following generic template as a learning tool and scorecard for each space.

For each S, every space to be addressed starts at Level 1. If improvements are noted for that particular S, then the appropriate higher level (with its accompanying score) is assigned in that column. The template allowed easy quantification so current levels of each of the 5Ss could be added together for a total score at the time the space was evaluated. Our plan was to establish an acceptable level (at least Level 3) for Sort in every eligible space before moving on to Straighten because we figured Levels 4 and 5 for Sort should follow soon enough. After Straighten was sufficiently in place (Level 3 or higher), then Shine would come next.

Beginning with Sort, every item in every eligible space was assessed by the following two criteria:

- Did we truly use it on a regular basis? If not, get rid of it. "Regular" was defined as something you would use within the next 2 weeks; if that wasn't likely, then get rid of it. We figured that if we really did need it later, we could get it soon enough.
- For those items of "regular" use, did we have too little, too much, or a reasonable amount/quantity for routine use? Determine how much should be in close proximity to where it's needed and no more, no less.

As we got going, the template for Sort was modified with more details and less abstraction so that employees could follow it more easily:

| Level | Expectation | Yes | No |
|---|---|---|---|
| 5b | A process for receiving feedback on maintaining necessary items is in place, reviewed, and followed up on during daily walks or team meetings | | |
| 5a | Evidence is present of further improvements focusing on elimination of unneeded materials and improving MRS[a] systems | | |
| **4** | ***"I know when materials are needed so I never run out"*** | | |
| 4f | It is clear where items that are required are stored or located | | |
| 4e | There is a clear process to respond to emergent or abnormal situations (i.e., out of stock or holiday orders) | | |
| 4d | It is clear when the quantity system is maintained, including expiration dates | | |
| 4c | Documentation is available that supports how the quantity system is maintained; it is clear who maintains the quantity system | | |
| 4b | There is Standard Work that defines what the quantity limits are (e.g., Kanban, Visual Controls) | | |
| 4a | All existing Visual Controls are clear, instructional, and effective for anyone to perform the work | | |

| Level | Expectation | Yes | No |
|---|---|---|---|
| *3* | *"I know what is needed"* | | |
| 3c | MRS levels have been established in Maximum/Minimum format only | | |
| 3b | All outdated/unnecessary signs and items have been removed from the workplace | | |
| 3a | All unnecessary furnishings and fixtures have been removed from the workplace | | |
| *2* | *"There are no unnecessary items"* | | |
| 2c | Are collection areas (e.g., shelves, cabinets) for non-essential items eliminated? | | |
| 2b | Is the area free of excessive inventory not needed to perform Standard Work? | | |
| 2a | Are all unnecessary items identified (Pink Tagged) for removal and less than 2 weeks old or inside of project scope? | | |
| *1* | *"I see unnecessary items"* | | |

*Notes:* aMRS *stands for "Materials Replenishment System," a common descriptor for how supplies are restocked.*

5S walk-throughs were conducted 4 days a week, starting at 10:00 a.m. and announced in advance, to evaluate each facility or administrative area at least twice monthly. The manager responsible for the facility or administrative space would host senior staff and others from the management team (composed of facility managers, technical program managers, middle level administrators, and veterinarians) for an hour. Every eligible space would be visited and then given a score (Level) as defined in the 5S template, to be posted outside the room next to the previous score, to indicate if we were making progress or not at that particular location.

How hard was it? It took us 20 months(!) to get 90% of the eligible spaces to at least Level 3 just for Sort. When we challenged staff about the necessity of each item in every space, it quickly became evident that folks were holding on to an excess of commonly used supplies so they would never run out, or were keeping an item because it was needed years ago and who knows when it might be needed again? As we pressed on, these habits of stockpiling lots of stuff around eventually unearthed the fears on which these habits were based. The notion of not having something at hand when it would be needed was considered by some as failing in their job, even after frequent reassurances that senior management would take the fall if we were ever truly out of a critical item and the program suffered as a result. And the fear of failing in that context far outweighed the fear of failing a 1S evaluation, at least at first.[7] The personal tension between not letting go of old stuff and wanting to be seen as loyal to the 5S campaign resulted in high emotions (including tears) during at least one walk-through. Constantly drilling the 5S template into everyone's head on the management team, combined with regularly scheduled and formal evaluations of each space, was the only way we were going to change old habits for the better. We knew that some felt our approach was oppressive, silly, and too anal retentive but I was convinced there was no better (and certainly no easy) way to replace wasteful practices with something better. After the resistance to Sort was overcome, Straighten and Shine came much easier and faster, to everyone's relief.

How successful was it? We eliminated hundreds of items from our workspaces, emptied five rooms, and made restocking of supplies more obvious and easier so we would neither run out of items nor have too much on hand. We even discovered some fairly odd items hidden under piles or in closets and long forgotten, such as a Sony Betamax video camera and a small cement mixer (the original purpose for acquiring it remains a mystery to this day). A special space was created to hold a lot of the discarded stuff temporarily so that those unwanted items could be donated to researchers and other departments if they wanted them. If an item wasn't taken by anyone else over the course of several months and after multiple e-mail blasts to the MGH research community, then it was sent out with the trash (I had to convince our person manning this dispensary that he would still have a job after it was emptied). Consider these other outcomes:

- We reduced the setup time for a rodent housing room in one of our facilities from 24 min to only 6 min. That in turn freed up the equivalent of 3.6 FTE's over the course of a year to perform other tasks.
- On-boarding of new animal care personnel was accelerated because every room was neat and clean, and items were clearly labeled; the initial training time for a new technician was reduced by 50% for large animal and 30% for small animal vivaria.
- 2000 ft$^2$ of space that had been used for storage was switched to more value-added space in which we could either house more animals and generate additional per diems to cover some of the fixed space costs we were already carrying, or re-assigned to procedure space for investigators (the cost of which came off our books); the net financial benefit to the program approached $165,000 per year.
- Just as significant was the fact that our 5S campaign engaged the entire workforce, requiring not only their full participation but their understanding as to why we were doing this, while sharing the pain and successes with everyone.

In addition to hoping to make our workers' jobs easier, we were also driven by the essence of lean to eliminate anything not valued by the customer. The best way to determine what's valued or not is to ask customers. To that end, we initiated dissatisfaction surveys with the researchers who relied on our services. My feeling was that serving our customers was the sole reason our department existed, so wouldn't it be logical to find out what they thought about the tasks we performed on their behalf and what might be lacking? ("see Chapter 10")[8].

What else is possible when lean is applied to laboratory animal care? Other lean management tools, such as value stream maps, root-cause and herringbone analyses, A3 forms, kaizen events, 5-whys, Plan-Do-Check-Act, and visual controls can be just as helpful as 5S in improving operations in our industry.[9] Additional lean-related benefits in the two programs I have directed include the following:

- reducing required personal protective equipment (PPE) and disabling air showers in rodent barrier facilities that use micro-isolation caging (because the cage is the barrier), saving $12,000 per year in supplies plus saving time by not having to don unnecessary garb or wait in air showers prior to entry;
- replacing rodent sentinels with swabbing exhaust plenums of IVC racks for health surveillance of the colony via polymerase chain reaction (PCR) tests for excluded microorganisms, which avoids buying 1600 sentinel mice and rats per year in addition to saving time by not having technicians transfer soiled bedding to sentinel cages during cage changes;
- testing mouse imports themselves via PCR swabs on arrival rather than relying on quarantine sentinels, avoiding an additional 4–6 weeks of quarantine per import for investigators;
- having vendors rather than our personnel deliver and maintain stocks of supplies (e.g., food and bedding, cage wash chemicals, paper and plastic goods) inside the barrier, liberating 400 h/year for staff to perform other tasks;
- using unheated wash water and no detergents to wash rodent cages in specially modified tunnel washers, saving on energy and chemicals while making the work environment more comfortable;
- acidifying drinking water for bottles and automated watering lines, eliminating biofilms or fouled water so that bottles can go longer before replacing.

What other routine practices are reasonable targets for lean-inspired improvements? Here's one: when hazardous chemicals are injected into animals as part of a research protocol, those cages and the soiled bedding they contain are often labeled as dangerous and require special handling and disposal precautions. But if the amount injected is tiny, compared with whatever is excreted intact and then greatly diluted in the bedding, why are those cages and their contents still considered an increased occupational health risk? Chemotherapy drugs are a representative example. They're certainly toxic, which is often the basis of their efficacy. But excreta from cancer patients who receive frequent drug infusions aren't considered toxic to the rest of us and don't require special capture and disposal. So what's different about mice in a cage injected with the same drug at likely the same dosage per unit body weight or body surface area? The answer is absolutely nothing, and programs shouldn't be burdened with the additional time and materials needed to manage those cages differently and having to incinerate their bedding.[10]

Along the same line of reasoning, protocols sometimes involve injecting unadulterated (i.e., not further genetically engineered) bacteria into laboratory animals to study disease and evaluate novel ways of detecting and treating infections. Federal guidelines assign Risk Group-2 (RG-2) and Biosafety Level-2 containment (BSL-2) for many bacteria that may cause serious disease in susceptible humans. Some of the same bacteria can be found in or on healthy humans and laboratory animals, the latter of which are housed and handled under Biosafety Level-1 (BSL-1, the lowest risk) containment if those bacteria are already present "naturally" in the colony. Examples include *Klebsiella oxytoca, K. pneumoniae, Pseudomonas aeruginosa*, and methicillin-sensitive *Staphylococcus aureus*. Consequently, laboratory animals with identical species of bacteria may be assigned BSL-1 containment if these microbes are considered commensals versus BSL-2 containment just because they're labeled RG-2 and administered to an animal for an experiment. And these two containment scenarios can occur in adjacent rooms in the same facility, or even in adjacent cages on the same rack. In both cases, personnel handling either type of rodent cage and its inhabitants wear PPE that meet or exceed requirements for working with these bacteria in a BSL-2 laboratory. But insisting on BSL-2 animal housing for these bacteria results in additional PPE and special disposal requirements that creates an illogical increase in time, money, and environmental pollution. Fortunately, the dominant US reference document for such things provides guidance on how microbes may be assigned to risk groups based on a variety of factors besides their pathogenic potential in vulnerable persons.[11] Another federal regulatory document that prescribes safety measures for using pathogenic microbes in animal research also assigns each of the four bacteria listed above to RG-2 but defers to the first (dominant) reference document for "information on agent risk assessment."[12] Thus, BSL-1 containment is allowed, at least at the federal level, for housing and managing laboratory animals after their inoculation with at least these four bacteria. How many institutions are taking advantage of this regulatory performance standard at no additional risk to occupational safety or animal welfare?

By this point I hope you're at least intrigued, if not inspired, about applying lean principles and tools to your program. If so, my advice is to start small, practice a lot, and engage those closest to the action. What activities or habits could be fruitful targets? Just ask your front line technicians what are they doing as part of their daily routine that seems stupid? If they trust you, they will name plenty of tasks that are puzzling if not irritating. Invite them to design alternative ways of doing something better, which could mean anything from easier physical tasks, fewer mistakes, better animal welfare or occupational safety, improved employee morale and fulfillment, to achieving lower costs. Then conduct a few trials that generate numbers rather than opinions, review the data with them, and discuss which option may represent an improvement. After several rounds of this approach, then one can introduce more sophisticated tools and lean vocabulary on a program-wide scale.

Another bit of advice: change of any sort is difficult for many to digest and program leaders should temper their ambitions when altering routines. Lean offers a good cautionary lesson of this. When someone at a higher level of authority tells the workforce that everyone is now expected to help eliminate unnecessary work, don't be surprised that many will hear, instead, that they're supposed to help eliminate unnecessary *workers*. The good news is that the reality is just the opposite, that getting rid of wasteful activities usually increases job security because employees will finally be able to do other, and often more important, things that weren't possible earlier because no one had the time. Adopting lean to eliminate waste can improve both the value and the affordability of a program without reducing headcount to cut costs. However, convincing everyone in the program of this requires frequent and enthusiastic reinforcement, accompanied by credible evidence that the work is truly getting easier and that no one's job is in jeopardy as a result.

Another caveat when undertaking lean is not to get distracted by the tools themselves. For example, checklists have become popular as a means to simplify processes and eliminate unnecessary work or mistakes.[13] But some managers focus too much on such means while at the same time failing to communicate sufficiently with all the participants. Consequently, something as simple as a checklist can fail to succeed if it's employed in an information vacuum with insufficient monitoring of other process components.[14] Moving to something more complicated than checklists, I became enamored with statistical process control (SPC) as a tracking tool for six sigma quality control systems[15] not long after my arrival at MGH. In order to achieve six sigma (six standard deviations from a target mean outcome, approximating one error or defect per one million events or items), one has to measure output constantly and apply those measurements to established tolerances. I thought that if we could track our throughput and rework and then share those findings with the entire department, it would identify which inefficiencies or mistakes were frequent versus rare, big versus small, etc. And because SPC generated numbers rather than personal testimony, any conclusions would be more palatable because they weren't based on subjective criteria. I quietly hired a consultant to look into how SPC could be applied to vivarium operations. We analyzed various elements to see what may be useful, including processes for animal husbandry, animal health, research services, finance, administration, and human resources. For tasks believed to be critical or representative of the program's quality profile as a whole, the consultant calculated current means and standard deviations, along with proposed upper and lower control limits for acceptable efficiencies or rework. We generated some spiffy charts and graphs, just like the ones in the SPC textbooks, and were almost ready to share these with the department when I realized how damaging it would be for us at that time. Rather than embrace the data as a measurement of the work, our workers would interpret these charts as personal critiques and morale would plummet. We weren't ready for such steely eyed tracking of output. Furthermore, it wasn't clear when we ever would be since there was no crisis of quality to fix and, in the absence of a crisis, such tools would be difficult to employ without harming the supportive culture we were trying to build. So the project was shelved. That's not to say operational error rates shouldn't be quantified for the purpose of improvement, but one needs to tread cautiously.

In retrospect, I gained additional insight from this experience based on the potential for conflict between six sigma and lean. Six sigma strives to reduce errors by performing the same task often enough to get it right every time, whereas lean is constantly on the hunt to eliminate waste which, in turn, may make (new) errors more likely as processes are modified. Both approaches can improve productivity and quality but they're quite different in their underlying philosophies. So whenever I hear the composite phrase "lean-six sigma" to describe a quality management strategy, it grates like the sound of fingernails drawn across a chalkboard. There is also misunderstanding about what either approach can or can't do, such as the perception that lean confines an organization to incremental rather than breakthrough progress.[16] This conclusion is more a result of limited thinking rather than limitations of lean. Six sigma and lean are quite compatible when both are leveraged in a sustained and enlightened manner, evidenced by the success of Toyota and other organizations in finding a constructive balance between repetition and change.[17]

I hope this brief overview of lean vivarium management convinces the reader that the title of this chapter is accurate. If done well over a long enough period of time, lean can result in any of the following outcomes.

*More time* to do the other things your team should have been doing but couldn't because they didn't have any time to spare.

*More space* to expand or diversify program activities without capital spending, or to turn over to the institution for other uses and thereby eliminate its fixed costs from your budget.

*More money* to return to the customer as reduced per diem rates or reduced program subsidies, to invest in new technologies or new processes, to perform original research on laboratory animal biology or medicine, to reward staff for their efforts and accomplishments.

*More employee fulfillment* to learn new skills, to participate in continuous improvement, to advance into positions newly created as lean initiatives identify unmet needs.

*More engaged program management* to understand what really goes on at cageside and determine more accurately what's essential versus what can be discarded or replaced with something better.

Perhaps lean is best exemplified by the story about three persons looking at a drinking glass filled halfway with water. The first describes the glass as half-full while the second describes it as half-empty. The third, a lean practitioner, asks why is the glass twice as large as it needs to be?

# Chapter 12

# Magnets and Glue

Laboratory animal care is, at its core, a service business. We don't grow, manufacture, or assemble something that can be stored on a shelf for later shipment to a middleman or end user. Instead, we create value in real time as we perform our tasks. The same applies to restaurants, airlines, insurance companies, supermarkets, and the like—while they rely on the products of other companies, it's the service experience that's the basis of the enterprise and attracts customers. Just like other service businesses, the primary assets on which we rely aren't patents or machines but employees, from office staff to cage-front technicians and everyone in between. Since it is highly unlikely most of those employees will be replaced by automation anytime soon, we need to reconcile two converging yet conflicting macro-trends if we want to sustain a dedicated and high-quality workforce on whom laboratory animals, scientists, and society depend all day, every day.

The first trend is the shift of hands-on animal care from a manual laborer job to a knowledge worker job. Beginning decades ago, animal husbandry positions were filled with the lowest cost employee possible because the tasks were simple, physical, and highly repetitive. Those were the days when cages were more primitive and often hand-washed, animal sources and their health status more variable, regulations less complicated, and science less sophisticated. Consequently, many animal care staff only needed a high school diploma, if that, and rudimentary fluency in the English language. Not much job-related knowledge or certification was required because one's supervisor would decide the work to be done and then closely monitor the workers to make sure it was completed.

Contrast that to today's programs, with their wide variety of animal housing, food and drinking water options, advanced cage sanitation/sterilization equipment, and expensive environmental controls. Add to that social housing, environmental enrichment, extensive colony health safeguards, subtle protocol endpoints sometimes driven only by a single gene or protein, and stringent compliance standards scrutinized by IACUCs as well as regulatory and accreditation agencies (with activists poised to denounce the smallest transgression). The vivarium is now a much more complicated workplace where lots could go wrong, resulting in serious and dangerous consequences for animal welfare, scientific data, occupational safety, and an institution's public reputation. In most programs, the typical manual laborer of yore has been rightly upgraded

Notes in the Category of C. https://doi.org/10.1016/B978-0-12-805070-5.00012-6

117

to personnel who are more educated, more knowledgeable, more certified, and given more responsibilities while working under less supervision.[1]

A good example of this shift is how both my previous and current programs re-assigned some tasks to animal technicians that historically were performed by veterinary technicians. In almost all programs earlier, and still practiced in many programs today, when an animal technician discovered a sick animal, a special card was left with the cage to serve as a visual signal to a veterinary technician about this health concern. Or if the problem was extreme in its severity or if many animals in the same room were suddenly affected, the veterinary technician would be notified immediately rather than wait for them to discover the cage card during routine rounds. Then the veterinary technician would observe the animal of interest, confirm the problem, and initiate a pre-established treatment plan or contact a staff veterinarian who in turn may then need to contact a researcher to discuss the animal's condition or likely fate.

While this process made sure a health concern was identified and addressed correctly, it also meant the animal was left possibly waiting in pain or distress until the entire sequence played out. For spontaneous health concerns for which diagnosis, prognosis, and likely outcomes were well established across the industry, we didn't believe a delay was justified before the animal was treated or euthanized to relieve its misery. Conditions in mice such as ulcerative dermatitis, dystocia, fight wounds, dental malocclusion, and preputial abscesses could be easily identified and, more importantly, immediately addressed by anyone with appropriate knowledge and using standardized treatment plans for each condition. So animal technicians were trained to detect, diagnose, document, notify, and initiate veterinary care themselves for common health concerns of mice and rats, rather than merely let somebody else know and move on to the next cage. Laminated charts with photographs of common health concerns now hang in every rodent housing room for quick and easy diagnosis.[2] Each health concern is described in simple language for mild, moderate, and severe levels of presentation. Each level is further detailed with its own treatment plan created by staff veterinarians (and highlighted in red if euthanasia must be performed soon). Those same treatment plans reside in a departmental e-mail library so the animal technician can quickly match the health concern and its degree of severity with the correct e-mail and send instructions to designated animal users for that IACUC protocol, in addition to marking the cage with a special red card. Facility supervisors and veterinary staff are copied on the e-mail for the record; adding other program staff to a given e-mail is a convenient option in case the researcher has questions, objections, or needs a little more time to collect one last and possibly critical sample from that animal. The e-mail likewise serves as a handy digital record that can be retrieved later for assurance purposes or collated with other alerts for retrospective data analyses.[3]

Some tweaking of the e-mail alerts was needed after this approach was launched at MGH. Its initial success was obvious, with a researcher response rate of 90% and an average response rate within 1 day.[4] Since adopting this approach for my current program, all health concerns are addressed the same day they're identified. In addition to standardized communications and treatment plans plus faster resolution, this new responsibility offered more variety to the animal technicians' workday and some career growth at lower labor costs to the department. If you're worried that this put veterinary technicians out of business, just the opposite occurred. Now that they were freed from these tasks, they could be re-assigned to new roles just as compatible with their training and job descriptions, such as rodent breeding services, mouse imports and exports, upping environmental enrichment, and providing training in and performing large and small animal anesthesia and post-operative care.

This way of managing common spontaneous rodent health concerns is described in detail only to serve as an illustration of how today's animal technicians can and are doing much more than they used to. The same goes for veterinary technicians who, in this case, were "bumped up" to higher responsibilities than they used to have, with similarly fulfilling experiences and opportunities for personal growth. This macro-trend in laboratory animal responsibilities is all well and good as long as there's a robust pipeline of new workers to replace those that leave the field or retire. How well we attract, recruit, hire, retain, and reward employees has a direct bearing on the quality and sustainability of the services we provide.

This brings us to the second macro-trend and how it could conflict with the first one in a big way. Consider these sobering demographic figures: when all the baby boomers will have reached the age of 65 by 2029, more than 20% of the total US population will be that age or older (vs. 14.5% in 2014).[5] What do these projections portend for laboratory animal care? Even as many of those baby boomers work beyond the age of 65 for whatever reason (e.g., remaining healthy and active at older ages, needing the income because of inadequate savings for retirement), it's unlikely they'll continue to be employed in our industry in significant numbers except at senior levels due to the physical nature of the work. Given our reliance on the US labor pool while it begins shrinking, what happens if we can't appeal to enough new workers and then hold on to them? What else can we do that we're not doing already? What other tactics and tools are out there for career attractants (magnets) and employee retention (glue) in our industry?

Lots of effort and money is being poured into primary and secondary school education in the subjects of science, technology, engineering, and mathematics (STEM). The hoped-for return on this investment is a better qualified workforce more likely to succeed in jobs that will require more STEM knowledge, while a side benefit will be a more STEM-literate society that can cope with the whirlwind of technological changes coming and can make better communal decisions on what we as a nation and as a planet want to do. The benefits of students better educated in science and mathematics are obvious for the biomedical research workforce and its segments that care for and use laboratory animals.

But it's those other students who are attracted to engineering and mechanical technology in whom I see unrealized opportunity for us.

Currently, jobs in cage wash, vivarium support operations, or whatever you want to call it are considered entry-level positions, where program employees often get their start. While you're earning your stripes in cage wash, you're expected/encouraged/being prepped to move into cageside animal care, from which later you may be promoted to a supervisory position or hired by a research laboratory (unless you're leaving to pursue a career ambition to get into graduate or professional school). Some folks remain in cage wash or get demoted to there if they can't qualify for or succeed at the "next" level. Others may move "back" to cage wash if they've developed allergies to animal dander but want to keep the job benefits and not have to relocate their family by taking a job far away. Or some may stay in cage wash if no higher openings are likely, or return there just for some workday variety.

But it shouldn't be this way, not anymore. Remember how cageside animal care has become more complicated and more fraught with risks if you don't know what you're doing and not doing it right every time? The same goes for cage wash—newer machines have more intricate features that are more dependent on computer software and sensors. More cage processing measurements are expected to document regulatory assurance of such things as wash water temperatures, autoclave cycles that truly kill microbes, and employing luminescent swabs to detect residual biological residues. Plus if you've switched your inventory management system to a just-in-time approach, any disruptions in cage processing now get amplified downstream for animal housing turnover. In other words, if you're no longer keeping as many supplies and spare caging around because you've had to reduce your expenses, there's less throughput elasticity when cage processing machines are off-line for maintenance or repairs. All of this means the job of washing/sterilizing caging is definitely more complex and scrutinized than ever before. So let's not continue to relegate cage wash employees to lower levels than cageside employees.

Instead, let's establish new specialties and career advancement tracks in operations support and materials throughput, emphasizing and embracing the mechanical and technological nature of the job. Let's seek students in vocational high schools and community colleges who want to learn how motorized things work. Kids with a passion for electronics, plumbing, machine tools, metal fabrication and joining, moving conditioned air through ducts, and other non-biological fields are ideal candidates to draw to our industry. They should be offered guided tours, student internships, and mentoring opportunities with skilled specialists to entice them to include our field as a rewarding career option.

I know what you're thinking, that there are at least two reasons why this is either nuts or totally unnecessary (or unaffordable). First, your institution already employs skilled tradesmen to install these machines, keep them running, and fix them when they fail, or pays manufacturers' service reps to do these sorts of things. So why do you need to invest the time and energy to duplicate those skill sets?

My riposte: that's all well and good but who's going to actually operate those machines every day? And if we continue to regard these activities as merely unstacking, lifting, stacking, and pushing stuff around while knowing who to call when things go wrong, we're short-changing our programs and our stakeholders. Second, these jobs will require higher wages than what you currently pay to wash and sterilize cages. So why, in an era of tightening budgets, would you want to shell out more money than you are now? Riposte number two: one gets what one pays for—a worker insufficiently knowledgeable or enthused about his or her work is less likely to identify ways to do that work better and less likely to help advance the program so it can get better in order to stay competitive.

After we attract and hire these people, what could be done for their career progression so they stay and don't leave for a job in a different industry? Remember, they were recruited and employed because they enjoy working with machines, not animals. So the conventional ladder of advancement in which cage-washers move into cageside jobs, often in parallel with obtaining the usual entry level of certification, won't do. Instead, I propose creating a new technician certification program similar to the popular AALAS version recognized as the industry standard for laboratory animal care technicians, at least in the US.[6] AALAS offers three certification titles corresponding to three levels of demonstrated knowledge about laboratory animal biology, husbandry, disease, use, and oversight. Its mirror image could involve levels of technician specialty certification for "vivarium operations" or "operations support services," and look something like this:

| Knowledge Level | Certification Title | Knowledge Level | AALAS Equivalent |
| --- | --- | --- | --- |
| 1 | Assistant Vivarium Operations Technician | 1 | Assistant Laboratory Animal Technician (ALAT) |
| 2 | Vivarium Operations Technician | 2 | Laboratory Animal Technician (LAT) |
| 3 | Vivarium Operations Technologist | 3 | Laboratory Animal Technologist (LATg) |

Just like the AALAS series, advancing to the next certification level would require studying relevant material selected by a panel of credentialed and knowledgeable experts familiar with laboratory animal facilities and equipment, and then passing a written examination on that didactic material. One could add a practical examination or problem-solving exercises for reaching various levels if these are deemed useful. The subject matter for each level could address the following needs and topics:

- Level 1 (maintaining a safe work environment, basics of sanitation and sterility): to include official safety warnings on machines; appropriate PPE to wear when performing common tasks; proper ergonomics for lifting, carrying, pushing, and pulling items in the course of performing the work; how detergents work; common means of sterilization (e.g., autoclaves, chemicals, heat, radiation) employed in the vivarium for caging, automatic drinking water manifolds, types of bedding and feed.

- Level 2 (how the machines work): to include common means of delivering heat, water, and chemicals to surfaces that need cleaning or sterilizing; how exposure time versus temperature interact; how various classes of microbes, especially pinworm eggs, bacterial spores, and non-enveloped viruses are eliminated by common methods of sanitation and sterilization (i.e., detachment, dilution, destruction); how to determine adequacy of cleaning or sterilizing equipment and processes; minimum cleaning outcomes required for conventional, specific-pathogen-free, gnotobiotic, and axenic animals; the pros and cons of static versus ventilated caging, or water bottles versus automatic watering systems; how IVCs and automatic watering systems work and how to operate them correctly; HVAC supply and functionality in the vivarium; applicable engineering and performance standards in the Guide and AWA.

- Level 3 (how to work smarter and serve one's internal "customers" better): to include concepts and tools for improving throughput efficiencies, such as value stream maps, takt times, and bottlenecks; managing just-in-time materials inventory and other supply chain variables; tracking and addressing quality assessments for throughput processes using statistics, probability, and analytics; how to conduct continuous improvement pilot projects that alter one variable at a time using established approaches such as "Plan-Do-Check-Act," "Define, Measure, Analyze, Design, and Verify," and A3 forms.

Advancing to the next level could be rewarded with a bonus, a pay raise, or a promotion, just like how some programs incentivize their animal technicians to attain higher levels of AALAS certification. Using this approach benefits vivarium operations/support staff by giving them an alternative career path with more stimulation and variety that fit their interests, as well as higher compensation. Employers benefit by retaining knowledge workers who can reliably and safely perform tasks of increasing complexity with less supervision, and who will stick around to help the program improve with respect to capital infrastructure and materials throughput.

More good news: we wouldn't have to create a recruiting pipeline from scratch. At the secondary education level, many vocational high schools throughout the country educate students in the basics on subjects that fit this need. For example, Massachusetts' vocational high school curriculum includes the following array of subjects, with the number of schools in the state that offer that particular subject in parentheses: Building and Property Maintenance (7 schools), Electricity (44), HVAC–Refrigeration (23), Plumbing (29), Sheet Metalworking (3), IT Support Services (26), Programming and Web Development (18), Electronics (17), Engineering Technology (16), Machine Tool Technology (32), Metal Fabrication and

Joining (36), and Robotics and Automation (5).[7] If you don't see anything here you like, get in touch with vocational education administrators and offer to help create one.

High school graduates too young or inexperienced for you? Let's move on to community colleges for slightly older job candidates looking to put their education literally to work. Again using my state merely as a case in point, all 15 of our community colleges offer credits toward their degrees for coursework completed at recognized vocational high schools in Massachusetts, so kids can get a jump start earlier, just like advanced placement courses in non-vocational high schools. These credits are available in 14 areas; the ones applicable to us include Information Technology, Business Technology, Manufacturing/Engineering, Machine Tool Technology, and HVAC.[8] So students interested in these subjects could be introduced to the laboratory animal industry while in vocational high school, encouraged to get an Associate Degree in the same or related subject at a community college, and enter our workforce at a higher education level. College graduates in their early twenties still not old enough or sufficiently seasoned for your program? Another business model of many community colleges is to offer non-credit training courses with certificates of completion for adult workers acquiring new skills in order to change careers or reenter the workforce. Massachusetts community colleges offer non-credit training courses relevant to our needs in Computers, Green Technologies, Manufacturing and Trades, and Science/Engineering/Technology, to name a few; presumably, you've got similar programs near your institution.[9]

In addition to the workforce becoming older, with fewer persons of younger ages entering the labor pool, there are other population trends at work on a national scale. By 2035, 23% of Americans will be Hispanic, up from 18% in 2015, and remaining, by far, the largest ethnic minority in the country. This change is even more pronounced at younger ages in that six of every 10 Hispanics in the United States were born in or after 1981. Combine these developments with a graying workforce, and it's evident that Hispanics will fill a growing need in the labor supply as the ratio of workers to retirees shrinks. Another looming trend: immigrants from all nations will comprise an increasing proportion of the US population, projected at a record 18% in 2065, versus 14% in 2015 and 5% in 1965. If you include the children of immigrants along with their parents, that entire group will account for 36% of our population in 2065, versus 26% in 2015. For the next 50 years between 2015 and 2065, the majority of population growth in the United States will be driven by new Asian immigrants (35%) and new Hispanic immigrants (25%).[10]

Perhaps not coincidentally, these elements also typify the history of our industry's US workforce. As I alluded to in the beginning of this chapter, employees whose native language is something other than English accounted for many entry-level hires in the old days. They were attractive as employees because they were eager and able to arrive on time, worked hard, appreciated getting the job if they didn't possess advanced degrees or credentialed expertise (so their options were limited), didn't contest the necessity of animal-based research and testing, and if you were looking for more employees, they knew plenty of family and friends back home who would be willing to come work alongside them.

And while the immigrant work ethic, on the whole, has been exemplary, an expectation has grown that foreign nationals know and use enough English in the vivarium. The old approach was to tolerate entry-level workers who had poor or no knowledge of English while relying on others of the same background to translate instructions and help make sure these workers did their work well enough. The lens through which this was viewed in those days placed a high value on cheap labor costs, absence of anti-animal research activism, and constant supervision, with hardly any change in a technician's daily routine. Yet failure to understand printed instructions or accurately document performed tasks can seriously threaten animal welfare or the reliability of experiments. Plus occupational safety is at risk if the worker is not able to follow instructions while operating potentially dangerous equipment or understand warning labels for chemicals and other hazardous materials. Thus, anyone working in our field in this country today should be expected to have an acceptable level of English fluency and literacy, both to be hired and to be retained. What should program leadership do to ensure there's acceptable comprehension of English by all employees? And how is that to be balanced with the looming reality that employers are going to have to out-compete each other to recruit US-born and immigrant Hispanics, Asians, and other minority groups to ensure they have enough employees?[11]

One starts with establishing an adequate baseline of English comprehension and communication amongst vivarium staff. We addressed this issue squarely at MGH after I arrived in 2002. At that time, 57 out of 65 animal care technicians were non-native English speakers of nine languages (Portuguese, Spanish, Russian, Cape Verdean Creole, Haitian Creole, Mandarin, Cantonese, Italian, French), and only one person was AALAS-certified. English literacy and fluency in some staff was poor, reflected in having difficulty in learning and performing new tasks. This also left these employees with limited or no possible advancement or career growth. MGH offered voluntary English-as-a-second-language (ESL) classes to all personnel after hours, but attendance was difficult if one was a single parent or working two jobs. The vocabulary that was taught was appropriate for daily life activities and working in a hospital (e.g., using maps, computers, ATM's, photocopiers, filling in job applications, reading job postings, personal fire safety, infection control) but not really germane to our workplace or jobs.

My department's strategic plan for the 3 years 2003–05 emphasized three objectives: (1) enhance individual staff competencies and increase AALAS certifications; (2) develop a closer rapport between department staff in the animal rooms and research staff conducting the studies; (3) improve productivity and standardize quality with new standard operating procedures (SOPs) and operations technology. Better English literacy was a key component to achieving these goals. So our department's ESL aims within that strategic plan were to develop an assessment tool for pre-screening future applicants; implement an assessment process to evaluate written and oral capabilities of current employees; offer all department staff ESL training via a customized course; assure vocabulary and language skills were adequate to understand and participate in vivarium operations improvements; offer re-assignment opportunities at MGH and beyond for those who were not successful in completing the course. We retained a professional ESL instructor to design and conduct English language learning and proficiency assessments, and consulted early and often with MGH Human Resources to make sure whatever we did would comply with hospital policies and applicable laws and regulations, and be deemed fair, transparent, and as supportive as possible.

All personnel whose native language was not English were screened by the instructor and placed in one of five groups for subsequent proficiency training (1 = lowest, 5 = highest and ALAT-ready). The initial assessment was a written test for all non-native English language speakers, including 30 multiple choice questions for grammar (e.g., "The rats are ____ the cage."), vocabulary ("Cages with litters will be cleaned twice a week. 'Litters' mean ____?"), and facility procedures ("What does PPE stand for?"), and a writing test ("describe your job; what do you do each day?"). The curriculum was based on a lexicon of over 320 words and phrases taken from workplace signage, cage cards, and other routine forms, SOPs, and the Guide. It was organized as Animal Species, Animal Anatomy, Verbs and Times, Cleaning, Physical Plant Maintenance, Cage Detail, Bedding, Animal Health, and Computers and Information Technology Tools. Examples of weekly course outlines are provided below:

*Week 1*
    Lexicon: Species, Anatomy, PPE
    Grammar: Verb Tenses
    Writing: Forms and Cards
*Week 4*
    Lexicon: Job Details, SOPs
    Grammar: Passive/Active, Adverbs of Time, Nouns
    Writing: Paragraph Parts
*Week 9*
    Lexicon: Animal Health
    Grammar: Listening and Text
    Writing: Writing reported speech
*Week 12*
    Lexicon: Vocabulary Review
    Grammar: Tense Review
    Writing: Writing resolution analysis and proposed resolutions

Class sizes were small (9–12 persons) and offered twice a week in 2-h sessions as part of employees' regular work schedule over a 16-month period.

Participation in the training sessions was voluntary, while successful passing of an English written and oral final examination administered by the instructor was mandatory if one wanted to remain in the department. The examination was given three times over the course of those 16 months, with the following results: 52% (30/57) successfully completed the program after the first session (~6 months), 74% (42/57) successfully completed the program after ~16 months, and 17 technicians became AALAS certified.[12]

If you're doing the math, the other outcome was that 15 employees failed to pass the examination after multiple attempts and, therefore, couldn't keep their jobs. It was a very painful parting because many of these folks had been working as animal technicians or cage washers for years, some with repeatedly glowing performance reviews (by whatever criteria had been used), and all of them were nice persons. Fortunately, MGH is a big place that emphasizes retaining employees whenever possible. So some of them were offered new jobs in the hospital more appropriate for their communications skills while others were provided generous outplacement assistance. It took the department some time to get over the trauma of that change but we were much better prepared to move forward because of it.

That was then and now is now. Regardless of the ESL proficiency classes we instituted at MGH, we never discouraged employees to speak languages other than English in the workplace. And for potential employees of Hispanic heritage who are not native-born but arrive in the United States speaking Spanish, if their English is good enough, there's no problem with them working in any level of laboratory animal care and medicine. But I'd like to push this a little further with a social/cultural hypothesis of sorts, that overtly sanctioning and even encouraging the use of Spanish in the vivarium by those who prefer to communicate to each other in that language will attract, keep, and advance more Hispanic employees in our industry.[13]

How to prove or disprove this hypothesis? For starters, we can make our field better known in predominately Hispanic communities and their schools, including career-day presentations in classrooms and vivarium tours, to be conducted in Spanish and preferably by Hispanic employees. We can also offer more local support for the teaching of STEM subjects. That may help spark students' curiosity to investigate what we have to offer. To incentivize and reward Hispanic technicians already employed in our industry, we should also re-energize the AALAS technician certification program in the Spanish language. Until recently, one could still buy a training manual from AALAS that was printed in Spanish for entry-level certification (*ALAT Manual de Entrenamiento*), but with the following caveat provided: "This manual does not include all of the topics that are included in the 2009 ALAT Training Manual and is not recommended as study material for the ALAT Technician Certification examination." That's because it was translated from the 1998 edition in English and hadn't been updated since. Making matters worse, there are no training manuals in Spanish for LAT and LATg certification levels, and all certification examinations are written only in English.

Given the strong representation of Hispanics in our industry in the United States, it shouldn't take too much arm twisting to assemble a group of bilingual volunteers to agree on a generic version of Spanish and translate all three training manuals and examinations from current English editions. Just to be thorough, a separate panel of bilingual experts would confirm the equivalency and accuracy of the translations. All of this could be followed by Spanish translations of the training manuals and examinations I proposed above for the new cadre of vivarium operations/support personnel. Beyond our nation's borders, AALAS could approach its sister Latin American organizations with an offer of free Global Partner memberships in return for their participation in a mutually beneficial campaign to recruit qualified Spanish-speaking workers for all of us.[14] These sister organizations also would be welcome to use the Spanish-language versions of training manuals and examinations, thereby further strengthening the AALAS brand beyond the United States. Such an initiative could leverage family and community connections of Hispanic employees here and in those countries for potentially even better outcomes.

Appropriate metrics before and after launch would be needed to confirm or refute my hypothesis that Spanish language initiatives actually draw in and keep more US workers in our industry in order to avoid a labor shortage over the next 20–30 years. If successful, no one would ever ask again "*¿Donde están los manuales y exámenes de entrenamiento en español para las certificaciones técnicas de AALAS?*"

# Chapter 13

# How Do YOU Define Quality?

What makes a good quality program of laboratory animal care and medicine, and how does it differ from a great one? How can you know for sure? Why does it even matter? According to the conventional wisdom, "good" usually is defined with respect to regulations and accreditation standards; one's program only needs to maintain close enough compliance to federal (plus state and municipal, where applicable) regulations, and close enough alignment with accreditation standards. "Close enough" is just what it implies, burdened with at least three drawbacks.

First, compliance and accreditation worthiness are judged during inspections by a USDA veterinarian that occur unannounced roughly every year, and once every 3 years by an accreditation site visit team of two or more experts scheduled months in advance, respectively. Because there are lots of moving parts in a vivarium every day, including when animals get sick or machines break and before they're discovered and tended to, such brief glimpses by these outside visitors may miss or identify problems that could be long-standing and truly represent significant gaps in program management. Or if problems are identified in the course of an inspection or site visit, was it just bad timing since the problems would have been corrected promptly regardless of who saw them first? Inspectors and site visitors are obligated to document any problem observed, and most will also note if it was quickly and satisfactorily resolved. But in the case of USDA inspections, it's still on the record as a non-compliance issue, and therefore fodder for activists looking for any reason to trash a program's reliability, no matter how trivial the issue (by contrast and thankfully, AAALAC findings and outcomes are confidential because accreditation is voluntary and not subject to public disclosure unless the institution so chooses[1]). There is also a chance that significant problems are missed if the inspection or site visit doesn't cover every room or the program is not forthcoming in self-identifying those problems. The odds of that happening are lower with site visits even though site visitors don't evaluate every animal room or every protocol. That's because the emphasis on accreditation lies on how well the program is self-monitoring and self-correcting with respect to established standards, rather than making sure that oversight of every animal in every cage is perfect all the time.

I served as a so-called "ad hoc" AAALAC site visitor for 24 years and hosted many AAALAC site visits, too. After getting a few under your belt, either as site visitor or host, you can determine fairly quickly if the institution that's being evaluated is worthy of reaccreditation or what needs fixing. In addition to determining

Notes in the Category of C. https://doi.org/10.1016/B978-0-12-805070-5.00013-8

how well the host institution complies with the Guide and other relevant standards and applicable laws and regulations, an AAALAC site visit allows a glimpse at the culture of the parent institution in addition to the animal care and use program itself. Are senior executives or administrators cognizant of the importance of laboratory animal welfare or are they paying it lip service? Does the condition of the vivarium's physical plant and capital equipment reflect their professed commitment to do the right things and supply the necessary capital or not? Do the scientists who use animals for their research or assays have good rapport with the veterinarians and IACUC office staff or are the parties in conflict, even if they try to hide it during the site visit? Are the animal care technicians knowledgeable and encouraged to share any concerns with their supervisors, or instructed to follow orders and keep their mouths shut? The answers to these and other questions are much more telling about the state of the program than if every corridor appears neat and tidy, and every mop and bucket are labeled and in their assigned locations. Every site visit is supposed to include a lunch between site visitors and employees invited by the host but expected by AAALAC to include non-affiliated members of the IACUC. This encounter can unearth salient details not obvious while looking at vivarium walls and doorways or reviewing documents in an office removed from the action. Consequently, not every room or every animal has to be seen in order to reach an overall accurate assessment by AAALAC's standards.[2] Sure, a few cages may need to be repaired or replaced, some vials of recently expired drugs may be discovered and then immediately discarded, and researchers may need to be reminded not to recap hypodermic needles before tossing them in the sharps container. But if these findings are few and far between while the foundation of the program and the support it enjoys from the top and the investigators it serves appears sound, then it's likely that the institution will and should be reaccredited without much fuss. Conversely, if there are detectable fissures between the major program participants, those can be more symptomatic of dysfunction and risk to both animals and science than rusty pipes or cracked walls. The reality is that most site visitors, having read the voluminous program description provided by the host institution before their arrival, can reach an accurate conclusion about the state of an animal care and use program within the first few hours of walking around and talking to employees. The rest of the site visit is then often spent merely gathering more documentation to confirm their impressions and writing up those impressions and recommendations for the exit interview.

What about self-assessment by the IACUC's semi-annual inspections and program reviews and post-approval monitoring process? Wouldn't they be a complementary assessment tool to external inspections and site visits? Semi-annual evaluations and compliance lapses that are recorded by the IACUC should be included in its reports to senior executives/administrators and meeting minutes, all of which are reviewed by USDA veterinarians and referenced in the USDA's inspection report if a compliance lapse involving USDA-regulated species occurred outside the official inspection. But again, these internal evaluations are usually planned in advance or spotty in their frequency, lending themselves to the same possible misses or precise and prescriptive minutiae that may or may not truly characterize how well a program is running. This is not to suggest that institutions cover up their problems; to the contrary, it's been my experience that they're harder on themselves than outsiders would ever be, even when they know their conclusions will be audited and possibly publicized. The checklist that many institutions use for their semi-annual program review is taken from a template provided by OLAW that's entirely optional but convenient.[3] It comprises the standard categories that are easy to evaluate with Yes or No answers (hence, its popularity) but fails to evaluate the more subtle elements, such as attitude, dedication, teamwork, and mutual respect that are harder to score but just as critical to the success or failure of a program. And citing the non-compliance findings of these internal documents in USDA inspection reports doesn't capture all species used by a given institution if some of these animals fall outside of USDA's purview.

Fair enough, you say, but how about the requirement for an institution that receives NIH or other federal funding to self-report "serious" deviations from the Guide to OLAW? Wouldn't that provide a fuller picture of how good a program is, especially with respect to compliance? Notwithstanding the limitation that this applies only to institutions (and their animals) that get federal money, my answer is "it depends." OLAW's threshold for institutional self-reporting to it is communicated by examples as well as general principles, supported by additional tables and flow charts, with the intent to provide more clarity and better guidance for what must, should, or may trigger official notification.[4] The only problem is that not every potential situation is covered nor is it ever possible to do so, as OLAW rightly and freely admits. For cases that aren't specifically described or borderline, a consult is appropriate. OLAW encourages institutions to phone in their questions first to see if an event or finding is to be self-reported.

All well and good except when inconsistencies and gaps remain (and always will), either in policy or logic. When representatives of OLAW are confronted at the podium of conferences and workshops with an especially vexing scenario regarding compliance and self-reporting, they fittingly instruct the questioner and the audience to use professional judgment in determining what should be done. Sound advice as long as the institution's professional judgment, presumably encompassing its veterinarians and IACUC, agrees with OLAW's. But what if it doesn't? What official recourse does the institution then have? Here's where it gets interesting. The AWA includes explicit provisions and a process for appeals, a common stipulation in regulatory law, should the registrant (research facility) object to USDA's enforcement decision or penalties.[5] Even AAALAC, though its accreditation is voluntary rather than compulsory, has a defined appeals process should a unit object to AAALAC Council's decision.[6] Yet there's no established or published process to adjudicate differences of "professional judgment" with OLAW.[7]

Please don't misinterpret my remarks as regulation-bashing. The use of animals in research, testing, and education needs effective vigilance to ensure the boundaries of allowed behavior are not exceeded. The public rightly expects an independent entity to monitor our behavior and mandate corrections when needed. Because the decisions of regulators and accreditors are scrutinized and sometimes publicly derided by the opposition, it's critical that the scientific community continue to participate in establishing and policing those boundaries to help prevent their violation and erosion. The combination of regulatory oversight and voluntary accreditation has served the field and the public well over the past 30+ years when it comes to complying with those laws, regulations, and standards. But as just described, simply complying doesn't tell the whole story because it can never capture the entire profile of any program (that's why "close enough" has been deemed an acceptable definition of program quality all these years).

This brings us to the second major disconnect between compliance and quality. By way of introduction, take one of OLAW's examples of a deviation that does not normally require self-reporting: "infrequent incidents of drowning or near-drowning of rodents in cages when it is determined that the cause was water valves jammed with bedding (frequent problems of this nature, however, must be reported promptly along with corrective plans and schedules)." Whether or not one can establish what qualifies as frequent versus infrequent is tied up in the professional judgment conundrum described above (along with the illogic that infrequent deaths from drowning don't have to be reported but infrequent deaths from dehydration do when water valves get jammed with bedding but are shut off rather than stuck on). But this scenario also reflects a practical compromise that's inherent in how we do our business and is independent of OLAW's directives, i.e., a general tolerance of catastrophic consequences as long as they're confined to one or a few animals, especially if these animals are small, common, and not USDA-regulated (e.g., rats, mice, or birds) so they're not prone to greater disclosure through public inspection reports. It's impossible to prevent automatic watering valves from ever malfunctioning, and one hopes to detect these situations and intervene quickly enough so that animals don't drown. But accidents, occasionally fatal, are bound to occur. When they happen, as long as the intervention limits the number of deaths and includes other appropriate corrections, the program remains in full compliance and accredited. Thus, we've knowingly traded zero defects for "close enough," even though the same approach wouldn't be tolerated for airline crashes, free-falling elevators, or downed power lines. That's not to equate animal lives with human lives but it speaks to the disconnect between compliance and quality (no errors).

The third difference between these two frameworks is a consequence of the purpose of compliance to establish a base for research institutions to operate, i.e., a lowest common denominator. There is no distinction held between a program being compliant and being really compliant.[8] This in turn eliminates any obligation or incentive to push beyond the required minimum and pursue innovation that could further advance the lot of the animals, the scientists, and the program. Consequently, where's the driver for continuous improvement? Most successful companies in other sectors are at the top of their industries, in part, because they're always evaluating options and opportunities to get better. And unintended consequences can also drive change; there are plenty of stories in the business press about how accidents have led to innovation.[9] By contrast, we discourage any experimentation with tried and true methods, and settle for just following the rules the best we can even if these rules don't always keep pace with new insights into laboratory animal biology and disease modeling in animal subjects.

Let's say one accepts the invitation to move beyond compliance to define quality for a given program of laboratory animal care and use. Where would you begin? There are so many components to consider—just peruse the Table of Contents in the Guide for starters. Should you focus on basic husbandry (Section 3), veterinary medicine (Section 4), or programmatic elements such as IACUC responsibilities, training of animal care and research staff, and occupational health and safety (Section 2)? How about the condition of the physical plant, institutional security measures, and disaster response contingencies (Section 5)? After you decide where to focus your attention, what about all those other components important enough to be included in the AWA and Guide, the semi-annual program review checklist, and on AAALAC's scorecard when site visitors arrive at your door?

Since defining and driving quality in the vivarium are leadership issues, a good place to begin is with values and priorities. What are your personal values as a director or manager? What are your vivarium quality priorities? Is of what you're most proud a consequence of quality or convenience? Are you truly an exceptional manager or just lucky? Which problems or mistakes are worse than others? Should failure to administer or document post-operative analgesics for an animal be given the same weight as using expired drugs on it? Should employee absences due to workplace injuries be considered on par with animal deaths from husbandry mishaps? Is finding unlabeled chemicals in a procedure room equal to housing naturally social animals individually (if isolating those animals is not required by protocol or veterinary considerations)? In wrestling with these questions and more, my management priorities have always tried to be ranked in the following order of significance, from higher to lower (even though "lower" is still important and not to be ignored): (1) human welfare, (2) animal welfare, (3) institutional welfare, and (4) community and public engagement.[10] Taking these categories and their order as a starting point, here's how the components of a high-quality program might be framed:

1. Human welfare
   a. Every person's physical and personal needs for performing their jobs are provided and documented in an appropriate and timely manner
   b. A fulfilling work environment that values teamwork
   c. A learning environment to avoid repeating mistakes and explore process alternatives
2. Animal welfare
   a. Every animal procured from external sources arrives in good health and is subsequently protected from unwanted infection by appropriate and timely monitoring of the health of the entire colony
   b. Every animal's physical and emotional needs are provided and documented in an appropriate and timely manner, within limits allowed by the research objective that is approved by the IACUC (or its equivalent)

3. Institutional welfare
   a. Every minimum requirement for regulatory compliance is met and documented in an appropriate and timely manner
   b. Actual operating margins and capital expenditures are faithfully managed to their respective budgets, barring unforeseen and exceptional circumstances
   c. A culture dedicated to employee safety and respect, laboratory animal welfare, legal obligations, financial prudence, and policy and process rather than politics
4. Community and public engagement
   a. Participation in organizations and events dedicated to laboratory animal science, medicine, and welfare
   b. Sharing evidence-based knowledge and discoveries with colleagues through peer-reviewed publications and similar channels
   c. Informing lay persons about what we do and why

The next step would be to convert these objectives to operational targets that facilitate numerical tracking. Remember, if you can't measure it, you can't manage it. And relative characterizations such as "better" or "worse," "bigger" or "smaller," "slower" or "faster" aren't very informative when you're considering making a change in how your operations are organized or responding to a customer's request for something new. Quantitative results are preferred over qualitative ones because they not only allow you to compare differences with more precision but also provide credibility and make it easier to communicate the differences to others who (always) resist change even when they claim otherwise. So figure out how to convert program activities and outcomes into numbers that allow you to compare the "before" and "after," as both a tracking mechanism over various lengths of time and as a learning tool to see what effects a given change in a process may have.

One tenet of general management is that someone else has probably already done the same thing or something close enough for what you're seeking even if it's in a completely unrelated business. So you only need to learn about it and then try it out in your world rather than start entirely from scratch. Many companies and organizations outside of our industry routinely define and measure quality by the numbers. For manufacturers, this usually involves measuring defect rates in the products they make. Hence, the advent of six-sigma as a popular metric (defined as less than four errors per every million possible chances for that error to occur in production, assembly, shipping, etc.).[11] Six-sigma goes hand in hand with another tool known as statistical process control[12] that establishes tolerance ranges and measures differences in (quality) outcomes as a function of changes in the manufacturing process. Companies that provide a service rather than make things also define and quantify quality as a key to their success, including six sigma and statistical process control metrics. Hotels measure newly cleaned rooms against a list of requirements and analyze guest rebooking rates, airlines monitor on-time arrivals and lost baggage, military war games are tracked by senior officers for timing and precision of execution, and so on. Laboratory animal care is a service industry, too. What quality management models are out there to borrow for our needs?

How about cities? Talk about lots of moving parts! For starters, there's traffic and infrastructure, security and law enforcement, housing, school systems, municipal water treatment, entertainment and tourism, taxes, and, oh yes, don't forget compliance with municipal, state, and federal regulations. It doesn't seem remotely possible that wrestling with all the disparate components of a city could be reduced to a definition or numerical equation of how well it's being run. But that's exactly what's happening today in major US cities. Recently, mayors of Boston, New York, and Chicago have been tinkering with metrics to simplify monitoring how well their cities are operating.[13] In Boston, metrics are compiled from 24 separate inputs, such as percentage on-time street light repairs, number of library users, violent crime statistics, and fire department response time. These data and the others are condensed into a single number known as CityScore to indicate Boston's overall "health" for today, this week, this month, and this calendar quarter. CityScore is updated daily and displayed in the mayor's office and elsewhere. A CityScore of 1 means Boston is meeting specific targets or trending along historical levels. A CityScore of more than 1 means the city is bettering those targets or historical performances while a score of less than one means just the opposite, and it's a signal for someone on the mayor's team to investigate pronto. And because we're talking about public government, anyone can monitor CityScore online and see the same number the mayor and his staff are tracking, in real time, as excerpted below.[14]

**LEGEND**
< 1 (FOLLOW UP)
= 1 (MAINTAINING)
> 1 (EXCEEDING)

**CITY SCORE**

MAYOR MARTIN J. WALSH

LAST UPDATED 4/1/2016

| | DAY | WEEK | MONTH | QUARTER |
|---|---|---|---|---|
| STABBINGS (TREND) | 2.42 | 1.70 | 2.03 | 1.72 |
| LIBRARY USERS | 1.58 | 1.72 | 1.62 | 1.67 |
| MISSED TRASH ON-TIME % | 1.25 | 1.25 | 1.25 | 1.22 |
| PARKS MAINTENANCE ON-TIME % | 1.25 | | | |
| SIGN INSTALLATION ON-TIME % | 1.25 | 1.21 | 1.10 | 1.04 |
| SHOOTINGS (TREND) | 1.20 | 5.61 | 4.97 | 3.16 |
| PART I CRIMES | 1.18 | 1.37 | 1.33 | 1.38 |
| SIGNAL REPAIR ON-TIME % | 1.15 | 1.10 | 1.08 | |
| POTHOLE ON-TIME % | 1.10 | 1.07 | | |
| ON-TIME PERMIT REVIEWS | 1.07 | 1.19 | 1.17 | 1.13 |
| EMS RESPONSE TIME | 1.05 | 1.05 | 1.02 | 1.02 |
| GRAFFITI ON-TIME % | 1.02 | 1.15 | 1.18 | 1.20 |
| BFD RESPONSE TIME | | | | |
| STREETLIGHT ON-TIME % | | | 1.02 | |
| BFD INCIDENTS | | 1.10 | 1.16 | 1.30 |
| EMS INCIDENTS | | 1.02 | 1.05 | 1.08 |
| 311 CALL CENTER PERFORMANCE | | | | |
| CONSTITUENT SATISFACTION SURVEYS | | | | |
| BPS ATTENDANCE | | | | |
| HOMICIDES (TREND) | | 2.87 | 4.23 | 4.39 |
| TREE MAINTENANCE ON-TIME % | | 1.25 | 1.18 | 1.18 |
| | 1.18 | 1.45 | 1.47 | 1.38 |

A vivarium can be viewed as a miniature city of sorts, with the program director serving as the mayor. We manage infrastructure (lighting, heating, hallways), housing (cages), residents (animals), visitors (researchers), employees (animal care providers and custodial staff), medical centers (veterinary resources), trash removal (soiled bedding), drinking water treatment (hyperchlorination or acidification), operating and capital budgets, regulatory compliance, and more. All of these elements compete for our attention and decision making on a daily basis. So let's try our hand at creating a version of CityScore that I'll call "VivariumValue" to capture measurable outputs from the myriad of routine activities that occur in a given animal facility every day. Using CityScore as a guide, what's to be displayed are not the absolute numbers but a summary of changes in these data over time, i.e., are things on an acceptable path in general, getting better, or getting worse?[15]

To that end, I've added to each of the above objectives some hypothetical examples in italics that can be measured and expressed as a (sample and arbitrary) baseline value or subsequent numerical change:

1. Human welfare
   a. Every person's physical and personal needs for performing their jobs are provided and documented in an appropriate and timely manner
      i. *Sample occupational safety metric: the incidence of accidents and injuries in the workplace is 1 per 10,000 person-hours of time spent in the vivarium, comprising animal care staff, researchers, facility maintenance employees, and occasional visitors (note that this metric is more demanding and informative from a workplace quality perspective, than merely tracking scheduled health and safety training or audits)*
      ii. *Sample occupational health metric: 5% of the required uniforms and PPE are damaged or excessively soiled on any given day*
      iii. *Sample personal needs metric: animal care technicians perform (only) eight weekend or holiday overtime shifts per year, in addition to their regular 40-h work week*
   b. A respectful workplace that values teamwork
      i. *Sample teamwork metric: new animal care technicians are cross-trained in all facilities and for all species within the first 6 months of the start of their employment, to ensure adequate coverage in the event of extended staff absences*
   c. A learning environment to avoid repeating mistakes and explore process alternatives
      i. *Sample learning metric: no more than three and no fewer than one departmental continuous improvement projects are ongoing at any one time*
2. Animal welfare
   a. Every animal procured from external sources arrives in good health and is subsequently protected from unwanted infection by appropriate and timely monitoring of the health of the entire colony
      i. *Sample preventive medicine metric: outbreaks of infection by excluded pathogens are eliminated within 30 days of their discovery*

b. Every animal's physical and emotional needs are provided and documented in an appropriate and timely manner, within limits allowed by the research objective that is approved by the IACUC (or its equivalent)

   i. *Sample animal physical welfare metric: the incidence of flooded rodent cages maintained on automatic watering systems is 4 per every 10,000 cage-days*

   ii. *Sample animal emotional welfare metric: the incidence of animal injuries in co-housed NHPs or rabbits is 1 per every 500 cage-days*

3. Institutional welfare

   a. Every minimum requirement for regulatory compliance is met and documented in an appropriate and timely manner

      i. *Sample regulatory compliance metric: the total number of minor and significant findings in a calendar quarter from semi-annual inspections and post-approval monitoring audits are six and two, respectively*

   b. Actual operating margins and capital expenditures are faithfully managed to their respective budgets, barring truly justified circumstances

      i. *Sample financial metric: the centralized animal care program is within 5% of its budgeted deliverable (revenues minus direct operating expenses), as tracked via monthly forecasts*

   c. A culture dedicated to employee safety and respect, laboratory animal welfare, legal obligations, financial prudence, and policy and process over politics and drama

      i. *Sample institutional responsibility metric: reported concerns about animal welfare are investigated, resolved, and communicated to regulators (if and as required) within 30 days*

4. Community and public engagement

   a. Membership and participation in organizations and events dedicated to laboratory animal science, medicine, and welfare

      i. *Sample participation metric: 90% of animal care technicians are members of the local AALAS branch*

   b. Sharing evidence-based knowledge and discoveries with colleagues through peer-reviewed publications and similar channels

      i. *Sample knowledge dissemination metric: one abstract is in preparation or submitted for presentation at a regional or national meeting per calendar quarter*

   c. Informing lay persons about what we do and why

      i. *Sample public outreach metric: one presentation about the care and use of laboratory animals is made to local science teachers and classes every calendar quarter*

Each of the above objectives is expressed as a quantity (incidence, percentage, unit of time, etc.) that's simple to measure and easy to tell if changes have occurred since the last measurement. And each of these metrics is amenable to investigation if values worsen or if one wants to raise the bar to a higher quality level. Many other objectives and metrics are possible in measuring how well a vivarium is functioning, and each program's individual characteristics will determine which ones are the most useful. But rather than try to define every element as a quality objective, select those that can encompass other activities in order to keep the list a reasonable length. Objectives and metrics can also be selected because they offer insight into how they may be related to inform a larger picture of quality. For example, CityScore includes student attendance but no other education metrics from the Boston Public School system, such as standardized test scores and graduation rates; one presumes that attendance is sufficient for this purpose in addition to being socially linked to library users (more=better) and graffiti (more=worse). Furthermore, quality objectives and metrics don't have to remain constant.[16] Introduction or removal of a specific class of animal models or species, type of caging, or sanitation process could trigger modifying how a particular program defines or measures quality. But it's important to remember that quality is determined by outcomes, not effort. So if manual cage washing is replaced by a robotic system, a clean cage is still the goal, regardless of how it's cleaned.[17]

The approach described above for tracking quality requires lots of measuring, lots of analysis, and lots of education and involvement of all parties that influence quality. But that's what a program of laboratory animal care and use that's dedicated to excellence should be doing, anyway. Where on an appropriate quality spectrum does YOUR program stand? Before you can address that question in a meaningful way, answer this one first: how do YOU define quality?

Chapter 14

# The Rise and Demise of Standardized Husbandry

*It's never too late to be what you might have been.*

George Eliot (pen name of Mary Ann Evans)

Henry Ford's genius was not in manufacturing automobiles but in expanding the market to make them affordable to the middle class. He accomplished that by dramatically lowering the purchase price while still being able to make money. That in turn was made possible by simplifying and standardizing the assembly process at such a low cost. The Model T was the means by which the Ford Motor Company came to dominate the automobile industry for two decades. It began after Ford learned how to produce and incorporate lighter weight vanadium steel into his cars, thereby enabling his fundamental vision as he recounted in his auto-biography: "The less complex an article, the easier it is to make, the cheaper it may be sold, and therefore the greater number may be sold." He then surprised his salesmen by declaring "in the future we were going to build only one model, that the model was going to be 'Model T', and that the chassis would be exactly the same for all cars." To extend that point, Ford informed them that "Any customer can have a car painted any colour that he wants so long as it is black."[1] Ford's assembly-line processes dropped the price of touring cars so much that over 15 million Model T's were eventually produced and accounted for 60% of all new automobiles sold in the United States in 1921.[2]

Laboratory animal care programs continue to swear by Ford's business precept of providing animal users only one type or model of cage, one kind of bedding, one defined item or assortment of materials for environmental enrichment, one means of delivering drinking water, and one set of macro-environmental parameters (i.e., lighting, light schedule, temperature, relative humidity, number of air changes per hour). This is a legacy from decades of steadily improving the quality of care animals, i.e., by reducing the variability of caging, feed, bedding, drinking water, detergents, etc. while gradually eliminating most pathogens from animal breeders and research colonies. Narrowing the variables not only eliminated lots of uncertainty but also offered two other advantages in managing programs. First, buying anything in multiples is often accompanied by volume discounts, so one pays less per unit if more units are purchased. And that's fine as long as one truly can use all those units. But I've seen plenty of vivaria with unused, and sometimes never unwrapped, capital equipment in storage that was part of a large procurement package driven by a volume

Notes in the Category of C. https://doi.org/10.1016/B978-0-12-805070-5.00014-X

**139**

discount mentality, with hopes that eventually the program will expand enough to use all of those idle items. Second, less variety means simpler operational processes and less chance of making errors that in turn could injure animals or risk non-compliance. Stated another way, if my job is to wash and assemble rodent cages or place clean enrichment devices in NHP or swine enclosures, fewer choices means I'm not as likely to forget something or do it incorrectly. For all these reasons, almost all programs continue to impose a very limited assortment of supplies and accessories for animal husbandry on their customers.

Such a stance means that occasional requests from scientists for animal housing components that don't match the house standard are usually denied or accommodated with reluctance. And there may even be a higher per diem rate charged, to boot, if the institution recovers a portion of its animal care costs by invoicing research grants or contracts. Or investigators may have to provide the non-standard items or perform the "irregular" husbandry tasks themselves, often without getting a break on per diem prices even though these prices include the very things they are now doing.[3] What if an investigator simply wants to replicate someone else's published findings? This is a common occurrence since science insists on independent confirmation of one's results. And given the recent crisis over irreproducibility in all branches of science, it's logical that as many components of that replicated experiment match the original version as closely as possible so any "noise" from confounding variables is avoided.[4] But what happens if the second scientist's vivarium doesn't have the same caging, food, bedding, environmental enrichment, water, or environmental parameters? A similar conflict occurs when someone moves her or his entire laboratory, including personnel, equipment, and samples, from institution A to institution B. If the new animal care program's husbandry components differ in any way from that of the old one, that principal investigator (PI) may be rightfully concerned that these differences may jeopardize the continuity of scientific data between the old place and the new one. On a larger scale, what if a PI's entire field is shifting to a new paradigm that requires unorthodox animal husbandry or housing? What if sand is found to be more appropriate than wood chips for hamster bedding, or ferrets are to be provided a foodstuff that differs from today's norm for a particular disease model? Despite these and other reasonable scenarios, the common stance of program directors is that this is our established package of husbandry components and it's not (practical or possible—choose one) to duplicate your or anyone else's setup if it differs from ours.

Such inflexibility has consequences. There are plenty of examples of how previously considered "trivial" differences in an animal's immediate environment may influence biological data as experimental endpoints becomes more subtle or simply because we learn more about various influences on animals every day. Not so long ago, it was shown that repeated washing and autoclaving of mouse cages made of some but not other commonly used plastics released a chemical that disrupted how egg cells divide in female mice housed in those cages.[5] More recently, the effect of different bedding materials on rats' sleep or mouse nest building behavior has been described.[6] Why should a program resist accommodating reasonable requests to deviate from the default husbandry package after findings such as these become relevant to an investigator's research?

Another consequence of inflexibility is the burden it may place on other programs less able to uphold the status quo. Almost all programs in the United States use rodent bedding made from wood or corn cobs. These materials are popular not only because they're functional but also because they're abundant and affordable in this country. But such ubiquity has become misunderstood as superiority to other materials, and preached around the world as a "best practice." But what if a program in a less wealthy nation at a subtropical latitude can't afford to import ground corncob or aspen wood chips for its rodent cages? It should be okay for that program to do the same thing we did in the United States decades ago, i.e., look around for plentiful raw materials that are byproducts of local agriculture or industry and see what may work just as well. Countries that grow large quantities of rice may consider using rice hulls for rodent bedding. One evaluation found rice hulls to be less absorbent for this purpose, meaning that cages would likely have to be changed more often than conventional bedding substrates.[7] But if labor is cheaper in these countries where rice hulls are an obvious option, then it shouldn't be ruled out. We risk making science more expensive in many parts of the world when local programs in wealthy nations demand particular materials and practices that in turn get elevated to international norms.

But must program leadership remain loyal to the concept of one-size-must-fit-all? Is providing the same components the best and only way to stay on budget and avoid unintended mixing and matching? The answer to both questions is no. One of our two central missions is to serve society by serving science (the other mission also serves society, by intending to spare every animal from unnecessary pain or distress in the course of routine care). Thus, we're obligated to accommodate the needs of our customers as those needs evolve. Extending that credo a little further, programs should invite its customers to voice their needs or curiosities about husbandry setups other than the standard combination. The good news is that alternative management strategies already exist in other industries that we can easily adopt in order to provide a greater variety of customer choices without breaking the bank or committing errors.

Since we're well versed in the concept of comparative medicine, we need to consider the notion of comparative *management*, i.e., how our operations may resemble those in other industries and learn from successful companies engaged in entirely different businesses. Lots of enterprises offer a great deal more variety of products or services than previously thought possible. Starting with Henry Ford's company, it now produces 27 models of cars, passenger vans, and pickup trucks under the Ford marque for just the North American market, each of which comes with a variety of available features comprising engine size, type of transmission, color, upholstery, etc. Imagine the complexity involved in designing, planning, manufacturing, assembling, and delivering these vehicles, each with a correct account of its features, while remaining profitable. Or let's consider the fast food sector—a menu at a McDonald's restaurant today offers 20 breakfast entrées, 7 different hamburgers, 5 chicken or fish items, 5 salads, 5 French fries, snacks, or other sides, 14 milkshakes or desserts, and 27 beverages. Ordering one from each category generates over 6.6 million possible combinations (and an admittedly unlikely feast), not including the various sizes of servings! Yet by the time you drive up to the pick-up window after placing your order, you only have to wait a few minutes at most for it to be ready. In addition, it's usually accurate and affordable, and it has been prepared by workers whose native language isn't always English. How does McDonald's bridge a seemingly daunting chasm between possibilities and actualities, all the while offering speedy and reliable service to its customers and financial returns to its shareholders?

The solution involves at least three parts. First, Ford, McDonald's, and almost every other company have broken down every process to its components and identified the ones that are shared between products (and therefore can be performed uniformly and in larger batches) versus the ones that aren't. French fries are all made from potatoes and prepared by McDonald's the same way—the only difference is the size of the container they're served in. Second, those commonalities enable companies to eliminate unnecessary steps from each process in order to avoid wasting time, money, and space. Third, by tracking the variety of output over time, a company knows which combinations are more common than others under differing circumstances and will manage its inventory and throughput accordingly. Just because Dunkin Donuts offers eight different kinds of bagels doesn't mean it bakes equal numbers of every kind every day for every store. Patterns of consumption can vary by geography, business location, time of year, time of day, changes in weather, and other factors. Bagel sales are matched to each of these factors with extensive data tracking to optimize the number of bagels delivered every morning so that outlets aren't left with too many bagels that have only a limited shelf life while making sure most customers aren't disappointed if their favorite bagel isn't available. This general concept of offering more variety to one's customers as a business strategy targeting individual tastes has even been given the catchy, albeit oxymoronic, name "mass customization."[8]

How may we engage in mass customization for routine laboratory animal care? In addition to the three parts mentioned above, two others are required that are commonplace in other industries but still lacking in ours: (1) leveraging lean management principles to eliminate unnecessary work and space, and (2) digitizing the vivarium. Both elements make it possible to provide a wider variety of services reliably and within one's budget. More to the point, expanding the array of husbandry options without employing lean management and digital tools is sure to create an unmanageable and unaffordable outcome (that is, unless you already have an obscene excess of idle labor, space, and money and are looking for ways to apply them). With the absence of these two important additions to the mix, it will be difficult to create the requisite space, time, and funds to accomodate all the unforeseen moving parts that may appear when investigators finally are invited to choose from amongst a variety of husbandry options. Employees will be tripping over each other and losing or forgetting critical details as labor and storage costs mount in attempts to keep everything organized and avoid mistakes. I'll describe below the inexpensive digital technology we have embraced in our program for this (see Chapter 11 for a discussion of lean).

As soon as I arrived in my current program, we began preparing to switch from an insistence on a single rodent housing package to an open menu of conventional options. The goal was not only to cater to our customers' preferences but also to offer anything within reason at the same per diem price and at no increase in our institutional subsidy. That meant, as hinted at above, reducing operating costs and liberating workers from pointless work sooner in order to offset any increase in costs and work that may accompany customized orders later. The first step was to introduce our staff to the principle of continuous improvement, to be applied to anything and everything we were doing. In order to make things more favorable for them, I started asking (and still do) that, from their perspective, what are we doing that's stupid? What makes no sense? What's potentially hazardous to animals, dangerous to workers, or just a big waste of time and effort? When replies started to emerge, we addressed them together in an open and collaborative fashion, and began piloting alternative ways of doing things. Little by little, folks began loosening up and coming forward with complaints, concerns, observations, and even suggestions to change what we were doing for something possibly better. The objective was not only to make their work easier but to get them less uncomfortable with the notion of change itself.

During this same period, a menu and ordering form were created on our department's internal website that researchers could access with their smart phones.[9] An early version of the mouse husbandry menu included these items:

*Caging*
Default—Allentown ventilated transparent plastic shoebox mouse cage
Available Options—Allentown ventilated transparent plastic shoebox rat cage—low profile (for mice), Animal Care Systems Optimice ventilated plastic mouse cage
(Don't see it? Please ask!)

*Bedding*
Default—1/4" ground corn cob
Available Options—hardwood chip, hardwood shavings, 1/8" ground corn cob
(Don't see it? Please ask!)

*Food*
Default—IsoPro RMH 3000 irradiated rodent pellets
Available Options—PicoLab Mouse Diet 20 (5058) irradiated mouse breeder pellets
(Don't see it? Please ask!)

*Water*
Default—Reverse-osmosis, chlorinated, continuous via automatic watering system
Available Options—clear or translucent red plastic water bottle, stainless steel sipper
(Don't see it? Please ask!)

*Environmental Enrichment Materials*
Default—2" square cotton Nestlet and translucent red plastic shelter
Available Options—Enviro-dri paper fibers, carefresh softwood paper fibers, BlockParty plastic tunnels to connect adjacent Optimice mouse cages, irradiated sunflower seeds
(Don't see it? Please ask!)

At the time this menu was created, these were the only items we carried in-house. But if someone wants another vendor's caging, bedding, food, water delivery device, or enrichment material, all they have to do is ask. That doesn't mean we'll go out and buy new cages and racks or large bags of the material requested. We have plenty of other and thriftier options to obtain these things, such as borrowing, renting, leasing, swapping, buying used, or asking the PI to purchase the item if it's only for her or his research and unlikely to be requested by other PI's anytime soon. And if a researcher only wanted to try out an alternative husbandry combination to generate some preliminary data, vendors will usually supply sample quantities, or we could ask our local colleagues for some and offer to return the favor at a later time.

The two last pieces of our customer-driven customized rodent care service are being finalized at the time of this writing. First, we need to make sure that every cage setup that differs from our default "recipe" will be replicated accurately so clean replacement setups are identical to the original combination ordered by the researcher. The conventional options were non-starters. Cage recipe cards were out because we already use lots of cage cards for other purposes; adding one more was sure to confuse and possibly get lost. Clipping plastic tags on cages wasn't attractive because the tags could become unclipped; clipping plastic tags on the cage's slot in the rack wasn't reliable because cages often are shifted to other slots and if the tag doesn't move with it, you're sunk. The hoped-for solution we're trying now are printable labels that would display a Quick Response (QR) code corresponding to that cage order and stuck on the side of the cage but with a twist—they're dissolvable and come off during cage washing, permitting the cage not to be permanently linked to that recipe. After each washing, cage setups to be made for new customized recipes can receive new adhesive labels printed in the same room where the customized setups are assembled (our "kitchen") before delivery to animal housing rooms.

The second missing piece is a suitably attractive retail display of caging and materials available for order, even those we don't carry in-house. Glass shelves in shiny white cabinets with back lighting will have transparent containers with lids holding different kinds of bedding, enrichment materials, and food pellets. On other shelves will be examples of various rodent cages and water bottles, even ones we don't stock. Each item can be picked up and handled by the customer for closer inspection and comparison shopping if desired. Each item will be accompanied by a label that includes the item's name, matching the one in the e-menu, and perhaps a brief description or veterinary advice regarding what circumstances the item should be used or not. Display cards will also have QR codes printed on them that when scanned with a smart phone will link to the supplier's website in case the shopper wants to review technical specifications such as the type of plastic used for the cages or bottles, or a proximate analysis of the nutrients in a given food.[10]

For each selected item, the menu will then ask the customer to indicate how many are desired, when they need to be delivered, to which room, and for which protocol? Items can be ordered in specified combinations or à la carte. Each submitted order is first reviewed by an animal care supervisor to make sure the order makes sense and, if so, it will be forwarded to the kitchen; if not, the investigator will be contacted to clarify the order before it's approved. In the kitchen (assigned the more mundane title of "cage prep room"), two copies of the order are printed before the order is assembled. Both copies go into a clear plastic sleeve that accompanies the order to its destination; one copy will remain in the animal room as long as that combination for that customer and that protocol is active while the other copy will accompany soiled cages to cage wash and be returned to the kitchen for preparing clean replacement cages with identical ingredients or features; that copy will then come back up to the housing room to remain until the next soiled cage(s) of that recipe are to be replaced.

How variable could the array of husbandry combinations become? The most complicated outcome possible (or "worst case," in the minds of some) is for every rack of cages in a given animal housing room is from a different company, and every cage on every rack differs to some degree from every other cage on that rack. This is the extreme scenario that skeptics may envision as they categorically reject even the possibility of offering such services in an economically sustainable fashion. But how much variety really is likely after inviting researchers to customize husbandry at no additional cost? Will it unleash a torrent of orders, each one differing from all the others? The answer is no. In fact, I expect there will be hardly any difference from what's already in use in one's vivarium. That's because if scientists don't want to alter their animals' husbandry parameters, the last thing they need is any change to the animals' immediate environment and diet; anything different from past norms could jeopardize the continuity of their baseline data and negative controls. So the vast majority of cage setups and husbandry components will remain unchanged for established customers. But offering husbandry options for the same price rather than imposing little or no choice is still the right thing to do.

There's also a more subtle influence minimizing the likelihood of animal husbandry variety running amok, based on how humans make choices. The conventional wisdom, at least in the United States, has been that offering more options is better because we prefer variety and cherish the individuality supposedly afforded by making choices for and by ourselves. This isn't a trivial matter, especially with respect to economics and succeeding in business. Companies large and small spend lots of time and money thinking about ways to expand the number of products or services they sell. They're convinced that current and future customers will be more likely to buy something that's offered if there are more selections offered. But does an expanded array of choices lead to more purchases? Sheena Iyengar, the S. T. Lee Professor of Business at Columbia University, describes "the choice overload problem" about how consumers choose which product to buy as the variety of available products grows.[11] Her research indicates the opposite of what conventional wisdom tells us, that consumers feel more stymied and are less likely to purchase anything as the array of options grows. A telling example was a study she conducted with a smaller or larger variety of flavors of jams on display at the entrance of a supermarket. If six flavors of jellies were offered for taste testing to customers as they walked in, 40% of persons stopped to try at least one flavor; if 24 flavors were displayed, 60% stopped at the table. That would seem to support the notion that more choices is better. However, only 3% of customers who stopped at the 24 flavors table actually bought a jar of jam, versus 30% of customers buying a jar when only six were displayed. The resulting math says that customers were six times more likely to purchase a jar of jam when six rather than 24 flavors were offered. The take-home lesson from this and similar research is that too many choices discourages rather than encourages one to make a selection (purchase). In order to avoid the choice overload problem and enable a pleasing outcome, Iyengar recommends employing four tactics when offering products or services to customers: (1) cut—provide fewer rather than more choices; (2) concretize—present the choices in a physically or photographically accessible format instead of just a list; (3) categorize—organize the choices in larger groups of shared similarities rather than smaller but more numerous bunches; and (4) condition the customer for complexity by initially offering fewer choices in some product or service categories before introducing other categories with larger numbers of choices later.

Wise words by which to abide for laboratory animal husbandry choices, too. You may have noticed that our customized mouse husbandry menu doesn't list every possible cage manufacturer, rodent diet, bedding, enrichment material, or means of delivering drinking water. That's intentional so our customers won't be turned off or tuned out. Instead, we can convene workshops at some future time demonstrating other options and introduce those gradually as either additions or replacements on the menu. The menu also intentionally avoids listing different ambient temperatures, room lighting cycles, or the number of air changes per hour for the IVC rodent racks to avoid confusing or overwhelming investigators. But if they want a higher or lower air exchange rate in their ventilated rodent cages, we'd be happy to honor that request if they ask. Or if they are trying to replicate an experiment from another institution and want to recreate every detail, we'll contact our colleagues there to find out the air exchange rate of their IVC racks and adjust one or more of ours accordingly.

How smoothly will this all go? I have no idea. But I'm confident we'll enjoy learning what we didn't know at the time and make suitable adjustments as this adventure really gets going. Given the likelihood of few actual takers, combined with innate psychological resistance to too many choices, programs have little to lose by merely offering their customers more than only one housing package. We may even learn something that advances our knowledge about laboratory animal biology and animal models.

Part IV

# Laboratory Animal Ethics

# Chapter 15

# Public Honesty About Laboratory Animals

Animal research saves lives. Period. It saves human lives. It saves animal lives. It has done so in at least the past two centuries.[1] It will continue to do so for some time to come. Just as important, animal research expands our knowledge of our world and theirs, regardless of any immediate or intended applications to human or animal health and welfare.

Opposition to the use of captive live animals in invasive and sometimes lethal research, education, and testing comes in three categories. First, it's unnecessary because acceptable alternatives exist and the longer we rely on live animals, the longer we delay transitioning to non-animal tools and approaches. Second, it's inaccurate; there are boatloads of promising findings from animal experiments that don't pan out in human patients, which shouldn't be a shocker because (those other) animals aren't human. So it's no surprise that the scientific literature is bursting with discoveries that cured lots of mice of every disease imaginable but failed miserably in human clinical trials.[2] Third and most fundamental, intentionally harming any sentient creature is immoral regardless of any potential or actual benefits other creatures, such as us, may reap.

Let's consider claims from animal research more closely with those objections in mind. Do the absolute statements made in the first paragraph always hold true? Does animal research constantly deliver the goods without fail? Is animal research the only possible route to discovery and subsequent medical advances? The answer to all these questions is of course not, and no responsible spokesperson defending animal research has, to my knowledge, made such claims. Instead, the honest and appropriate position is that until we can confidently replace animals with alternatives that are scientifically reliable in all experiments in which animals are used today, it would be short-sighted and put patients and others at unnecessary risk to abandon animal research prematurely. But that's not as snappy a sound bite as "ANIMAL RESEARCH=TORTURE!", "END THEIR NEEDLESS SUFFERING NOW!", or other vitriol. So what we're left with are complex truths versus simple lies.

Notes in the Category of C. https://doi.org/10.1016/B978-0-12-805070-5.00015-1

Those truths represented in the opening paragraph are examined in some detail below, in no particular order, to understand their complexities better:

- Very nasty things, indeed, are occasionally done to research animals. That's because nasty things, such as serious and sometimes fatal disease and injury, happen to people and animals we love that we'd prefer to reverse or prevent. Scientists create and study models of disease and injury solely for the purpose of understanding how those diseases and injuries occur and how they may be overcome with better drugs, vaccines, medical devices, wound dressings, etc. If a particular feature of a disease or injury isn't modeled as closely as possible to the human condition of interest, then the outcome of the research may be less likely to be useful for that human condition. And the more severe, painful, grotesque, or heart-rending the condition is, sometimes (but, thankfully, not always) the animal model must be faithful to as many components of that condition as possible in order to be informative and justify the animal's use. So otherwise healthy animals can, indeed, be subjected to cancer, diabetes, congestive heart failure, broken bones, burn wounds, ulcerative colitis, Parkinson's disease, AIDS, and other terrible conditions. Even so, when you walk through any vivarium on any given day, over 90% of the animals appear normal and comfortable. Many of these 90%, especially if they are mice and zebrafish, are used only for breeding while others may be awaiting the intended effect that could appear days, weeks, months, or even years later; some are used in experiments that inflict no pain or distress at all. But there's no denying that likely painful or distressing things are being done to some of the remaining animals; many, but not all, of which are then provided analgesics or other means of relief to reduce their discomfort. Every year the Gallup organization asks Americans their views on hot button moral issues such as abortion and human cloning. Included in the annual survey are questions about the use of animals in medical research. In 2016, 53% of respondents said that "medical testing on animals" was morally acceptable while 41% thought it morally wrong. Another 4% thought it depended on the situation, and 2% had no opinion. Interestingly, the moral acceptability of medical testing on animals was lower than "buying and wearing clothing made of animal fur" (59%) or "medical research using stem cells obtained from human embryos" (60%).[3] Favorable attitudes about animal research are trending down (a 10-point drop over the past 10 years in this annual poll), whereas those opposed are rising at the same rate. In the past, there used to be a significant gap between age groups, with older respondents more likely to support animal research while younger respondents opposed it. What was intriguing was this gap was pretty consistent over the years of surveys, which meant that as persons aged, they appeared to switch from opposition to acceptance. I always thought that was because younger persons are in better health, on average, and don't need medical care for themselves or loved ones until they get older. It's when they became spouses, parents, grandparents, and eventually elderly (or more likely to be caregivers of their now elderly parents) is when they became more desirous of medical advances for themselves and their families, so their attitude toward animal research would change accordingly. But that no longer seems to be the case.

In 2015, Gallup conducted a poll specific to animal issues that showed one-third of Americans polled were "very concerned" about how animals used in research are treated, with another third "somewhat concerned." On a related question, 32% of respondents felt the statement, "animals deserve the exact same rights as people to be free of harm and exploitation" came closest to their view about the treatment of animals (up from 25% in 2008), whereas 62% most closely aligned with the view that "animals deserve some protection from harm and exploitation, but it's still appropriate to use them for the benefit of humans" (down from 72% in 2008). Thankfully, only 3% chose "animals don't need much protection from harm and exploitation since they are just animals" (unchanged).[4] The age gap that was evident in Gallup's prior annual moral issue polls evaporated in this animal issue survey on the question of applying human rights to animals: 31% of those aged 18–49 years were in favor, whereas 33% of those aged 50 years and older were also in favor.[5]

Given that background, how best to respond? Accompanying the societal permission granted to do occasional harmful things to laboratory animals are laws, regulations, and guidelines designed to limit to the extent possible the degree of pain or distress involved, avoid using more animals than absolutely necessary, and employ non-animal alternatives whenever appropriate (popularized as the 3Rs). So if we play by the rules, embrace the so-called guidelines so tightly that there's no room for interpretation or dissent, and can recite the 3Rs in our sleep, what's the problem? It's the mistaken belief that being in compliance is sufficient justification for doing these things to animals. And that's a big problem because logic never trumps emotion, and statistics never trump stories. Furthermore, human beings are just inherently wired to believe bad stories (e.g., harming laboratory animals) more likely than good ones (e.g., animal models contributing to cures).[6] Advertising companies know this in their pursuit of your emotional buttons, the same as patient advocacy groups, humane pet shelters, save-the-children NGOs, etc. So why would we ever think for a minute that a deep and widespread human affection for other animals combined with an unease about inflicted harm on animals could be countered by a defense that since "we follow the rules," we're in the right?

A good lesson in how playing the regulatory compliance card falls on deaf ears is the response of the Tufts University School of Veterinary Medicine to a group of its graduate students protesting a research study involving six laboratory dogs many years ago.[7] Each dog was to have osteotomies performed on both hind legs in order to evaluate different approaches to making fractures heal better. The dogs were then to be euthanized so the bone tissue could be examined in greater detail. It was the killing of the dogs at the end of the experiment that those students would not accept. The school's public defense was that the study was necessary and based on sound scientific principles. In addition, all applicable laws, regulations, and university policies were followed in reviewing and conducting this study. That was a poor rebuttal to the imagery conveyed by the experiment and did little to ease public concerns inflamed by press releases issued by the protesting students. But the school did not permit the media to interview the veterinary surgeons performing the research who, as animal care professionals, could have described this study as involving the same types of injuries that happen to pet dogs after they've been hit by cars, talked about the number of dogs with broken legs every year from such accidents, and how the knowledge gained from this research could result in better outcomes for owners' dogs with the same traumatic injuries.

To that last point, we're finally getting smarter at not relying solely on facts but including emotion to explain animal research and its value to the public and the media, as a counter-weight to the powerful emotional imagery used by activists. The Foundation for Biomedical Research, Americans for Medical Progress, and other animal research advocacy groups are experts at reminding everyone what's at stake if society bans or even impedes the legitimate and compassionate use of laboratory animals. Even the choice of words is important in relating how animal research continues to benefit patients and their families, by getting more familiar and less technical. Thus, our stories now talk about "babies" instead of "neonates," "teenagers" instead of "adolescents," "grandparents" instead of "the elderly." And stories have gotten personalized so the public is introduced to a real patient and his or her disease, not a data point. We finally stopped bringing a knife to a gun fight.

- Not all experiments work as planned. Sometimes they uncover a wrinkle about what we previously thought about a particular phenomenon of nature. Sometimes they point the way to an entirely new line of investigation. Sometimes they fail totally. And for any experiment that's never been done before, there's no way to predict its outcome. One should have an expectation, otherwise known as a hypothesis, that the experiment is designed to prove or disprove. But one actually still needs to perform that experiment and there's never a guarantee it will work as hoped. That's why the entire process is called research. First, you search and then you often have to "re-search," i.e., search again. And when you finally think you've got something worth publishing in a reputable journal to share with colleagues, you also understand that science depends on independent verification just as much as it depends on discovery. Thus, your experiments will be performed by others to see if the same results are obtained and if they fit with your conclusions. Sometimes they do, sometimes they don't.[8] Marc Kirschner , the founding chair of the Department of Systems Biology at Harvard Medical School, summed it as well as anyone several years ago when he wrote, "In science, faster, better, and cheaper are not as important as conceptual, novel, and careful."[9]

- Other truths related to the unpredictability of research come to mind when recounting one of the charges against animal research, that it's not even useful for studying human disease and injury since laboratory animals aren't human. Because of this fundamental mismatch, as the argument goes, we continue to waste billions of dollars and hundreds of thousands of animal lives under the erroneous premise that animals can model human biology closely enough. Two examples are often trotted out as convincing evidence of this position. The first is sepsis, when bacterial infections get into the bloodstream and cause a life-threatening cascade of irreversible changes in the body that are sometimes too late for antibiotics to overcome. Not only is this a serious medical problem, it's also a fairly common one, with estimates of 2% of all cases requiring hospitalization and 10% of all ICU admissions annually in the United States alone.[10] Drug after drug has failed in sepsis clinical trials over the past several decades despite pre-clinical data supportive enough in each case to allow testing these drugs in septic patients. All of these failures have served as a poster child of sorts to proclaim that all animal testing of new drugs is misleading and, by extension, immoral. The other common example given is the continuing failure of biomedical science to come up with a successful AIDS vaccine, again despite billions of dollars spent over the past 30 years studying simian immunodeficiency virus (SIV) in tens of thousands of macaques as well as other retroviruses in other animal models.

While these two examples are accurate as to their particular facts, the conclusions drawn about the futility of animal models conveniently avoid other larger truths. First, that science is hard, and biomedical progress is even harder. Perhaps we've been spoiled with so many medical successes over the past 100 years that we expect anyone in a white laboratory coat surrounded by test tubes and centrifuges can find a solution for any illness, and soon. Sepsis and HIV infection have proven so resistant to medical counter-measures because they are very difficult problems, indeed. Biomedical science, as an industry, and the lay media haven't helped to dampen expectations. Every discovery, it seems, comes with predictions of eventual health benefits, even though responsible scientists and journalists are sure to mention that lots more needs to be understood and overcome before said discovery will treat or cure patients.

The second related truth is that, while some illnesses, such as sepsis and HIV infection, remain intractable (for now), many others have been overcome with life-saving medicines as a direct result of animal models. HIV offers a good reminder of this. While we still continue to pursue an effective AIDS vaccine, persons in developed countries who are infected with HIV are no longer dying of AIDS by the thousands as they were only 30 years ago. That's because combinations of anti-viral drugs now keep HIV in check to avoid irreparable destruction of one's immune system. That achievement directly stemmed from studying SIV in macaques to understand how the virus and various anti-viral drugs interact in the body, an animal model that continues to yield discoveries at the time of this writing that could not have been possible by experimenting first in HIV-positive individuals or studying HIV in human cells in a dish.[11] And monkeys weren't the only animals that contributed to successful HIV therapy in the early years of the AIDS pandemic. One of the first effective anti-HIV drugs developed was azidothymidine, commonly known as AZT. Prior to its market approval, too many babies born to HIV-infected mothers became infected themselves; sometimes while transiting the birth canal, sometimes by ingesting breast milk during nursing. Using mouse retroviruses and lactating dams, scientists first learned that giving AZT to pregnant mothers could prevent retrovirus transmission to their newborns, a medical practice that quickly became mainstream.[12] So when the lack of an AIDS vaccine is touted as blanket evidence that animal models don't work simply because they're the "wrong" species, the truth is not so simple.

- Not only is science not easy, it's also not fool-proof. Sometimes accidents and mistakes happen. When laboratory animals are involved, sometimes these animals are harmed or killed unintentionally. Albeit rare, when accidents or mistakes do occur, they more commonly involve animal care rather than animal experimentation (purely because many of the animals in a given vivarium aren't undergoing experimentation on any given day, as explained above). Cages can break and expose sharp edges leading to injury; automatic watering valves can leak and flood cages to the point of drowning their inhabitants if not discovered early enough; fighting between cage mates can be missed during daily rounds and result in severe wounds or death; food can become spoiled and dangerous to eat if not stored properly; unwanted microbes can infect a colony, sometimes requiring depopulation even if the animals don't become sick—they may no longer be suitable for research due to more subtle changes caused by the infection, or the bug of concern may be particularly adept at evading eradication unless drastic measures are taken. Moving from animal care to animal use, drugs required by the research protocol can be accidentally administered at the wrong dose, via the wrong route, or to the wrong animal; surgical incisions can dehisce or become infected; vital signs may be taken improperly or not documented,[13] to name a few.

Some of these incidents can be avoided with better equipment, better training, or better oversight while others cannot. Fortunately, they're extremely infrequent although you wouldn't think so when reading the so-called summaries of official inspection findings and self-reported mishaps to federal oversight bodies that occasionally get trumpeted by animal activists. These reports sometimes lump minor findings together to give the impression of a "long" list of purportedly egregious errors as obvious evidence of gross negligence by the guilty institution. Any lapse in general housekeeping may be grounds for an inspection citation even if it has no direct impact on animal welfare. Yet when it appears in an inspection report available to the public, it can be added to a misleading tally of the "worst" animal research institutions. Like anything appearing in print or online, especially if involving an emotionally charged theme, one must read beyond the headlines knowledgeably to determine if the story is right. This is not to excuse or dismiss serious blunders, and one accidental animal death is one too many. But a healthy dose of perspective is advised. For instance, let's take flooded cages and accidental mouse drownings caused by faulty automatic watering systems. These systems are popular because they involve less labor and equipment than water bottles that have to be placed in every cage by hand and washed between every use. One advantage of a bottle is that it only contains so much water so if it leaks, so one is left with soggy bedding in the cage and sometimes soggy animals that need to be dried off and warmed up from the chill. By contrast, a leaking lixit valve connected to an automatic watering system won't stop. Consequently that cage continues to fill with water and its inhabitants can perish if not discovered in time. The leakage rate of automatic watering valves is usually lower than that for water bottles, but that's little consolation to researchers and the mice themselves if serious or lethal water leaks actually occur. But how common are flooded cages and resultant deaths? A couple of years ago I compiled data from several programs that shall remain unidentified, to compare the incidence of past cage flooding and mouse death episodes due to leaking automatic watering valves. At first pass, those numbers seem quite high when presented as follows:

| Source | Number of Flooded Cages | Number of Mouse Deaths |
|--------|------------------------|------------------------|
| A | 159 | 14 |
| B | 107 | 10 |
| C | 613 | 0 |
| D | 35 | 10 |

But whenever examining incidence data such as these, it's important to have additional details, such as how many total cages were in use during the time those flooding data were collected? A handy denominator I recommend is cage-days, i.e., the number of cages occupied multiplied by the number of days those cages were connected to automatic watering systems. So let's look at those same numbers but as numerators along with their corresponding cage-days as denominators[14]:

| Source | # of Cage-Days | # of Flooded Cages | Flooded Cage Rate | # of Mouse Deaths | Mouse Death Rate |
|--------|---------------|--------------------|--------------------|-------------------|------------------|
| A | 386,781 | 159 | 0.041% | 14 | 0.004% |
| B | 252,000 | 107 | 0.042% | 10 | 0.004% |
| C | 766,014 | 613 | 0.080% | 0 | 0.000% |
| D | 159,817 | 35 | 0.022% | 5 | 0.006% |

If you combine the above data, the average drowning mouse death rate from these four sources comes to just under 19 per million cage-days. But that doesn't take into account all the mice housed in those cages and therefore at theoretical risk during all those cage-days. Every cage involved in these four cases was the standard ventilated plastic shoebox cage commonly used today, with a maximum of five adult mice allowed per cage. An estimate I use to convert cage numbers to mouse numbers is an average of 3.5 mice per cage, including litters of young before they're weaned. Applying that estimate gives us a risk of drowning deaths at around five out of every million mice over the entire time span tracked.

As stated above, one accidental drowning is one too many, and zero defects is a worthy goal but not always possible or practical.

- Even rarer than accidents are instances of inexcusable neglect and callousness toward laboratory animals. Those who perform animal research come from just as wide a spectrum of human behavior as any other group so, inevitably, bad actors can appear. When their bad behavior comes to light, the glare of negative publicity is rightly swift and severe. However, outcomes after said publicity are sometimes more complicated. Take, for instance, the seemingly institutionalized disregard for NHPs at a large and well-funded research center some years ago based on a secret video that recorded multiple episodes of preventable injuries and likely deaths as well as avoidable pain and distress.[15] Whatever the veracity of the exposé and regardless of any penalties and corrective actions imposed, one was left in the dark about the genuine culture and values of the organization regarding laboratory animal compassion and care. If I were consulted (which I was not) after the news broke and after changes, if needed, were made, I would have insisted on inviting in the media for a tour to see whatever they wished and not only as a one-time event. An invitation would have been extended to responsible journalists and humane societies to come back any time for follow-up visits, with more public outreach thrown in for good measure. Nothing less would have repaired the reputational damage suffered by this institution even if the research they were conducting was commendable and the charges of neglect were groundless or contrived.

There's no denying that my approach in this case would have been radical and impossible for many on my side to swallow. But there's no social medicine as good as sunlight and fresh air, metaphorically speaking. And if one couldn't explain and defend what was going on at that institution regarding acceptable animal care and use, then there was no justification for it to remain in business. On a broader front, we too often shrink from opportunities to describe our work to lay audiences, and the resultant silence creates vacuums of information and opinion that opponents of animal research are only too happy to fill. And if these protectionist views and claims go uncontested, the public hears only one story line and has no choice but to believe them. But while ticking off the benefits of animal research, one also needs to acknowledge to the public that animal research involves making some animals sick or injured in order to save some lives. I'm not ignoring the reality that any of this invites verbal abuse or worse. And in an era of trigger-happy social media that offers anonymity, these attacks can be pretty scary.

Consequently and not a surprise, it's hard to find scientists willing to speak up and speak out. A common frustration of research advocacy groups is difficulty in finding scientists or their employers (1) disposed to speak about the value of animal research, and (2) able to explain their research in easily understood terms. A bad interview can be stuffed with highly technical jargon, perhaps attempting to show off the intelligence of the person being interviewed but failing miserably to connect with the audience. In past times, there was also too much of scientists implying in an interview that "I know more (better) than you, and because I'm important and saving lives, you'll just have to trust me and leave me to my research." Either approach just reinforces how arrogant and clueless the interviewee was. I've been fortunate to work in organizations that always mention pertinent animal research when announcing scientific breakthroughs, while at the same time acknowledging the ethical responsibility of using animals only when deemed necessary and in the least harmful manner possible. That's in stark contrast to other places where the word, "animal," literally can never be mentioned in any research news in order to avoid negative publicity, no matter how important laboratory animals may have been for whatever discovery or advance is being touted.

- Scientists are human, just like the rest of us. Sometimes they resist change, just like the rest of us. Thus, they may not easily let go of what gave them earlier success, such as particular animal models or reliance on animals in general, even though science is supposed to use the best new tools available at the time to make new discoveries. Nor are scientists always willing to give their mentees the intellectual freedom to explore new avenues of investigation. So, yes, it would not be surprising if animals are used in some laboratories more because of historical precedent and personal familiarity than because of the current state of the science. To that point, I've heard about younger researchers actually being discouraged by their mentors in abandoning animal-based approaches even when non-animal approaches are scientifically advantageous. A reluctance to change one's beliefs and habits isn't unique to scientists. There's a cynical adage with a modicum of truth that states "progress occurs one funeral at a time." This can apply to many communities besides ours.

- Science, if not individual scientists, is grounded on objectivity, transparency, and peer validation. If someone finds a more accurate method of probing a particular biological phenomenon, and it's confirmed by others, then it becomes mainstream. And adoption of that new method is even faster if it also costs less, takes less time, or involves less paperwork. Consequently, if a non-animal approach proves better for a particular scientific question, then animal approaches will be eventually replaced (even accounting for any laggards mentioned above). A good example is how monoclonal antibodies are produced. In the early days, one injected antibody-producing tumors into the abdomens of mice. As these tumors grew, the liquid they leaked into the abdomen was rich with antibodies secreted by those tumors and harvested multiple times by aspirating fluid from the abdomen. The mice were essentially living production vats of antibody but at significant cost to their welfare. These tumors weren't benign and clearly took a toll on their host as they grew. To address this, IACUCs imposed limits on the number of times (usually, two or three) the abdomen could be tapped before the animal had to be euthanized. Or if the mouse's health declined below a specified threshold, it was to be euthanized even if more abdominal taps could have been obtained. Fortunately, cell culture, with its simpler and more precise components, has replaced the mouse as the dominant means for monoclonal antibody production. Using an animal to produce antibody or any other protein is a relatively crude approach because other stuff besides tumor fluid is constantly produced in the course of simply being alive. So the antibodies extracted from the tumor-bearing mice still needed to be separated from lots of other molecules before they were useful research reagents. In addition, monoclonal antibodies, from the day they were announced as a new research tool also held much promise as a human drug. However, producing them in mice would be problematic for both patient safety and industrial economies of scale. So mice were swapped for cell culture because cell culture is better, even though mice were still an option. This is merely one of many examples of replacing laboratory animals with non-animal systems simply because the latter proved superior to the former once those replacements were perfected.

• In a related vein, many experimental approaches such as cell culture and biochemical reactions in vitro (literally, "in glass"), touted by activists as non-animal alternatives, still often rely on animals as sources for cells or chemicals. If scientists are studying how kidney cells change during various stages of renal failure, they may use kidneys taken from mice that can yield millions of cells for multiple experiments before moving on to studying human cells. The good news is that fewer animals are required if they are merely sources of cells or enzymes instead of individual experimental subjects themselves. This is especially true if the biological material extracted from the animal then becomes self-propagating, as with cultured cells that are transformed to never stop dividing. In any event, it's disingenuous or ignorant to include those particular approaches as examples of how research no longer needs animals at all.

• The vast majority of animal protection groups are populated by considerate, mature, knowledgeable persons, as opposed to the handy stereotype sometimes portrayed by our side of law-breaking radicals willing to commit violence. It's been heartening to see responsible animal protection groups condemn extremist tactics when property is destroyed or scientists are threatened with bodily injury. Furthermore, I've literally enjoyed thoughtful conversations with scientists, ethicists, and veterinarians dedicated to the immediate elimination of the use of animals in invasive research. And I've attended scientific conferences where ardent animal protectionists make cogent arguments, on the basis of objective data and statistically sound conclusions, for replacing animals with non-animal alternatives in toxicology and other disciplines (while at the same time, the organizations they represent simultaneously make ridiculous claims or worse). Make no mistake—we disagree on numerous fronts, and their pronouncements jeopardize the ability of scientists to make further medical progress with the best tools available today, including laboratory animals. But at the same time, it's embarrassing how disrespectful we sometimes are to these individuals in public forums when painting all of them with the same crude brush.[16]

There have been previous attempts at establishing dialogue between laboratory animal advocates and laboratory animal opponents, with a sincere willingness by both sides to engage. In each case, much was expected, if not promised, due to the prominence of the organizers at the time of launch. But suspicion and resistance from others in both the scientific and animal protection communities were immediate and predictable. Thus, those initiatives uniformly failed, mostly because they were too public too early, too vague in their goals, and they did not include sufficient representation of all stakeholders. But the pursuit of dialogue should not be abandoned because of past disappointments. What could be gained? There are many topics of mutual interest, such as promoting the 3Rs or adopting out laboratory animals to private homes, where opinions could be respectfully aired and received in pursuit of understanding. Such understanding in turn could lead to perhaps new approaches that, while neither wholly satisfying to both sides, may enable progress with respect to both science and animal welfare.

Given the polemics associated with animal research since its beginnings, is there any evidence that such an undertaking would have a snowball's chance of working? Lots of entities that oppose each other in other contexts engage in discussions, usually behind the scenes, to identify common ground and avoid mutually destructive outcomes. When nations do it, it's called diplomacy. When corporations do it, it's called strategy (and sometimes collusion, by regulators). One other contentious ethical arena fraught with life-and-death consequences and violent extremists offers at least one instructive experience in this regard, namely abortion. Six leading Boston-area women, three staunchly pro-life and three just as resolutely pro-choice, agreed to meet in secret after a crazed gunman in 1994 killed two persons working in local clinics where abortions were performed. Over the course of 5 years and under the aegis of the Public Conversations Project, a Boston-based organization that enables focused discussion over acrimonious issues, these participants gathered together as individuals and not as representatives of their respective organizations. Just as importantly, their objectives did not include common ground or compromise. Instead, they merely expressed their personal beliefs in a safe environment, with the assistance of professional facilitators, and in hopes of dialing down the invective to avoid more shootings. One outcome of their meetings was a remarkable joint public statement that defined each group's values and recounted their feelings and shared experience.[17] No one's stance was altered but that wasn't the point. In fact, their statement includes the observation that "While learning to treat each other with dignity and respect, we all have become firmer in our (respective) views about abortion." Nevertheless, bridges of communication and understanding were established that helped reduce vilifications of one side by the other.

Thankfully, there hasn't been human bloodshed around the topic of animal research in the US yet (although there have been a few close calls) that might have stimulated dialogue between the warring parties. But that's no reason to avoid off-the-record discussions with no agenda beyond just talking to rather than beyond each other. What values are shared? What personal background may drive someone's beliefs and motivations? Are there common areas of significance that could benefit from collaborative brainstorming away from the media and one's respective tribe? It would be great if there was a venue in which responsible and thoughtful adversaries could discuss animal research issues in confidence, without personal attacks or risk of disclosure. All participants would have to abide by complete confidentiality as well as 100% consensus if a position paper or opinion piece was ever issued. Under these rules, what's there to lose by not trying?

Chapter 16

# Increasing Reduction Options

One of the tenets of Russell and Burch's 1959 landmark book, The Principles of Humane Experimental Technique, is to minimize pain and distress simply by using fewer laboratory animals whenever possible. Specifically, this was defined by the authors as "reduction in the numbers of animals used to obtain information of a given amount and precision."[1] In the book and ever since, this goal was abbreviated as "Reduction," one of the iconic "3Rs" along with Replacement and Refinement (or, as I prefer, the active verbs: Replace, Reduce, Refine). As a result, the 3Rs remain deeply embedded in the language guiding various oversight bodies and the laboratory animal user community on what's considered ethical and acceptable with respect to using animals in research, testing, and education. External regulatory and accreditation entities require internal entities (IACUCs) to apply the 3Rs when reviewing protocols.[2–4] These protocols must provide convincing evidence to the IACUC that the author of the protocol has considered non-animal alternatives (Replace) but didn't find any that suited the aim of the proposed experiments. This is usually done by conducting a computerized search of the scientific literature, using key words pertinent to the animal procedures that are planned. The author must also describe the rationale for determining how many animals will be needed for each experiment, why that is the minimum number of animals needed (Reduce), and how endpoints that are expected to be painful or stressful are to be mitigated, if possible (Refine).

Recently, serious concerns have arisen over the growing failure of researchers to reproduce the results of others that are published in peer-reviewed scientific papers, including but not limited to experiments involving laboratory animals. In some cases, at least 70% of studies in animals have failed to yield the same results when repeated by other scientists, regardless of disease category, prestige of the journal in which the original paper was published, or the prestige of the institution where the research was conducted.[5] Another metric of the magnitude of this problem is the financial cost of animal-related studies that can't be faithfully reproduced, estimated at $28 billion a year in the United States alone.[6] Most of these findings have been traced to the original publication having faulty experimental design or selective inclusion of only those data favorable to the paper's conclusions rather than providing comprehensive data sets that may not be so supportive.[7] One of the components of poor experimental design is not having enough animals in each group of variables, i.e., the

Notes in the Category of C. https://doi.org/10.1016/B978-0-12-805070-5.00016-3

**163**

sample sizes are too small, to generate statistically sound conclusions that will stand up to inspection and attempted repeats.[8]

What could be the reasons for using fewer animals than necessary for reliable scientific outcomes? Certainly, one could simply be the cost. If obtaining research funding is difficult, as it certainly has been recently, scientists are incentivized to squeeze every dollar to try to accomplish the same with less, including but not limited to smaller numbers of animals. But could another reason arise from the past three decades of regulatory oversight, both intramural and external, that scrutinized every proposed animal experiment with the 3Rs echoing in everyone's mind? Could misinterpretation and overly aggressive pursuit of Reduction deter investigators from using enough animals, thereby jeopardizing the validity of the results of these experiments? Have we encouraged an oversight tyranny of sorts that has ironically, albeit unintentionally, resulted in wasting rather than conserving animals? A return to Russell and Burch's original language may be illuminating in this regard:

> For reduction purposes, as we have noted, the statistical method has a key property—it specifies the minimum number of animals needed for an experiment. This statement needs qualification. It certainly is always possible, in accordance with the arbitrary but workable concept of significance level, to decide after the event whether enough animals have been used. This saves needless repetition, and where, as sometimes in bioassay, workers are familiar with the amount of variation to be expected, a number found to give significant results can be fixed upon for regular practice. Exact treatments of the problem of choosing the right number in advance on the basis of experience are limited in scope so far.[9]

It's been pointed out that Russell and Burch did not equate Reduction with Minimization.[10] As is evident from the quoted passage above, they recognized the uncertainty inherent in performing first-time experiments and included a fitting willingness to accept the need for possibly more animals earlier so fewer animals may (or may not) be needed later. Nor did they consider Reduction as a worthy goal independent from the avoidance or diminution of pain or distress. To the contrary, the authors' moral basis for Reduction was explicitly established to prevent or reduce pain or distress. Misapplication by oversight bodies of Russell and Burch's original intent for Reduction while claiming this principle to justify the imposition of fewer animals in a given protocol could lead to experimental reproducibility problems if insufficient numbers of animals were approved.

Leaving aside how Reduction may be construed and imposed, there are experimental circumstances in which animal numbers have been reduced, sometimes dramatically. Take the advances in imaging that began over 15 years ago, for instance. Previously, if someone wanted to study what changes occurred inside an animal over time, such as how infection or cancer spreads or how a fractured bone heals, the usual approach was to euthanize small groups of animals (almost always rats or mice) at different time points in order to harvest their organs or tissues for microscopic analysis. That approach required enough animals to ensure there were consensus "snapshots" at each sampling interval for which the animal had to be euthanized to obtain the requisite tissue samples for evaluation. Nowadays we have rodent-sized imaging machines with the same technologies used on patients that enable one to follow the same small animal over the entire course of its illness and recovery. That eliminates the need for lots of rats and mice to provide post-mortem samples at each time point to piece together the whole story of what was going on inside of them.[11] And some of these medical scanning modalites have also been used for zebrafish, an even smaller species.[12]

A newer example of Reduction pertains to animal care rather than animal use.

Another research milieu in which some animal numbers are already reduced is regulatory toxicity or safety testing, driven by practicality as much as by fealty to the 3Rs. The FDA requires new drugs and vaccines be tested for toxicity in animals before human clinical trials can be approved. That's done in an attempt to identify the nature and severity of adverse side effects that patients may experience if the drug or vaccine is allowed to be administered to persons. Thus, the intent is to use doses that are high enough to cause detectable disease. A related endeavor is safety testing in animals, which is looking for evidence of harm at doses that would occur at accidental or occupational exposure levels, and driven by the EPA, Consumer Product Safety Commission, and other regulatory bodies charged with protecting us from receiving unintended doses of a given chemical or other potential threat (NB: the distinction between toxicity testing and safety testing is subtle but important with respect to the degree the animal's tolerance may be challenged). Most animal toxicity or safety tests require two mammalian species, one of which is almost always a rat or mouse.

How many animals are usually required for such tests? For rats or mice, the minimum number of animals per dose group is usually at least 10–20, determined in part by the specifics of that protocol; for larger species (most commonly, rabbits, dogs, or NHPs), that minimum number is usually 4–8.[13] Using fewer larger animals has been acceptable to regulatory agencies because of the much higher cost of procuring and housing these animals, as well as having to produce expensive test chemical or biologic in larger quantities to achieve the same dosages on a body weight basis as for the rodents.[14] This dichotomy between rodents and non-rodents always puzzled me, even when I was a study director responsible for conducting tests of this type years ago. I applaud the pragmatism of not requiring companies to pay for additional and more expensive large animals, and suspect public opinion may have some influence, too— there has been less mainstream political sensitivity about using hundreds of rats or mice versus hundreds of rabbits, dogs, or NHPs for testing a given drug or other chemical. But the contrast between what's an acceptable minimum number for rodents as opposed to other mammals begs the question: if fewer dogs or NHPs provide enough animals to reach scientifically sound conclusions about the toxicity or safety of a chemical or biologic, why doesn't the same apply to rodents? All of these are mammals, endowed with metabolic pathways and anatomy close enough to ours to serve as official stand-ins to detect possible hazards. Just because mice and rats are smaller and cheaper shouldn't be reasons for insisting they be more numerous in toxicity or safety testing protocols. Conversely (and even more important), if that many animal subjects are truly needed from each species for statistically sound conclusions, does it imply that higher, rather than lower, numbers of large species should be used? If so, is requiring fewer large animals risking the value of the results and, therefore, possibly using those animals in vain as well as jeopardizing our safety? As long as animal testing is required by regulators charged with protecting us from unknown or unreasonable exposure risks, is there an argument to be made for equalizing the number of animals needed, regardless of species, cost, or popular sentiment?[15]

Moving on from the existing applications of Reduction described above, there are a couple of other strategies that are less well known but warrant a fresh look. The first involves control groups. A canon of scientific study design is to include an untreated ("negative") control group with other groups subjected to varying doses of a virus or amounts of a drug where the effects are unknown. One then compares the results between all the groups to see if the results are statistically (truly) different between the known (control) outcome and the novel treatment, as well as to discern differences between various doses or time points. Every experiment almost always encompasses its own untreated control animals to ensure every animal in the study experiences the same environment, to make sure any differences are a result of only the intended experimental variable. If a negative control group isn't included in the experiment for comparison to treated groups, any resultant data are sure to be challenged and likely discounted when presented at a scientific symposium, described in a grant proposal, or submitted for publication in a peer-reviewed journal.

When negative controls are considered over years of time and across many laboratories and research institutions, that translates to lots of animals that aren't subjected to any treatments at all when used in experiments and testing assays. Accepting the fact that they serve a crucial purpose in scientific progress, is there a way their numbers could still be Reduced? In 2006, I was part of an international panel convened by ILAR to review what was then known about laboratory animal distress and then to issue a publication on the recognition and alleviation of that distress.[16] One outcome from that assignment was an invitation from SCAW to speak at their annual conference the following year. I had served on SCAW's Board of Trustees as its president over a decade earlier and was happy to oblige. The conference organizers wanted me to present the ILAR distress panel's findings and recommendations as well as offer some personal perspectives on avoiding or minimizing distress. One of my slides listed various ways to prevent distress, including using fewer animals, with the rationale that less aggregate distress would be experienced. On my list of examples was what I thought was a harmless suggestion, i.e., to share negative control animals amongst separate protocols and scientists. I thought this was reasonable as long as all the animals in the shared negative control group had nothing done to them until euthanasia. Or if something was going to be done to animals in the negative control group while they were still alive, such as taking weekly body weights, the same would also have been done to all the animals in all the experimental groups for all the protocols sharing those negative controls.

During the question and answer period following my prepared remarks, a scientist in the audience who was from an academic institution was quite irritated by this proposition and couldn't fathom why anyone would suggest such an unorthodox concept. I replied by presenting the following scenario: if all the protocols sharing these negative control animals occurred in the same room at the same time, were housed in the same type of caging and changed into clean cages the same way, provided the same food and water delivered by the same animal care staff, subjected to the same procedures as stipulated for negative controls in each of the pertinent protocols, and had sufficient blood or tissue samples to supply all those experiments, then why *not* share them since everything (*ad nauseum*) is the same? My skeptic remained unconvinced and said it would never work, or even if it did, it would never be acceptable under peer review. Perhaps there's a path of less resistance to shared negative controls through for-profit drug and chemical companies in performing their toxicity and safety assays since the incentive to save money on behalf of shareholders is omnipresent, compared with academic laboratories that operate under different pressures, and perhaps that strategy is already widespread in the for-profit sector. Either way, if sharing negative control groups became widespread, the annual savings in animals and dollars could be huge.

The second uncommon Reduction alternative I want to highlight is borrowed from human clinical trials. Whether it's a Phase I trial in healthy volunteers to assess safety, a Phase II trial in a small number of patients in a single hospital to assess efficacy initially, or multi-center Phase III trials involving hundreds or thousands of patients to confirm the efficacy demonstrated in Phase II, all of these trials conventionally enroll and test groups of persons at approximately the same time if not always in the same locale or region. This "batching" of human subjects provides a large-enough sample size to meet the regulators' need for rigorous statistical confirmation that the results are valid. It also provides the trial's sponsor with some economic efficiencies since data are being gathered for the same product at the same stage of evaluation and around the same time.

Although a highly organized process, the modern clinical trials paradigm is rooted in the fact that for thousands of years, physicians treated their patients by playing the averages. A medicine was usually selected because enough persons with the same condition had responded well enough to it previously. This practice was followed even though some patients showed no effect or, even worse, the medication made them sicker. And physicians knew that toxicity could occur, so regulations were established in the last century to require potential side effects to be clearly listed on the container of each prescription drug warning against this possibility. But as long as enough patients showed acceptable improvement and not too many of those treated experienced severe problems, that drug was considered potentially applicable to everyone with the same condition unless it needed to be relabeled or withdrawn by regulators.

Despite incredible advances in diagnosing illnesses and understanding the molecular basis of pharmacology, this wholesale approach to treating patients has remained unchanged. A drug is still prescribed for one person based on how others with the same disease or injury have fared. This practice continues despite acknowledging that not all patients with the same disease respond the same way and that there has been no way to predict a particular patient's reaction until he or she begins taking medication. At the same time, the cost of medical care is becoming an ever greater burden on organizations and entire nations, and paying for prescription drugs that don't work or make the patient worse is an expense that is fast becoming unaffordable, financially as well as ethically. But this is all changing.

New genes and gene interactions are being reported every month that are responsible for not only diseases but also for drug responses. Differences in the structure and activity of these genes can have a profound effect on how specific drug responses can vary between individuals. Soon we will be able to quickly evaluate every patient's DNA sequence and customize a medical treatment plan based on that person's unique genome. The result will be choosing a drug to address more precisely the patient's particular metabolic profile and avoid toxicities. Consequently, different patients with the same diagnosis may receive completely different prescriptions based on the person's individual genetic details. This conceptual advance has been popularly labeled as "precision" or "personalized" medicine and is getting a lot of attention.[17]

Making precision medicine a reality will require changes, or at least options, in how a new drug is tested before it's approved for market. One such option that is being championed is known as "N-of-1" trials.[18] Rather than administering a given drug to groups of patients for the purpose of seeing whether or not the average response at an (average) optimum dose is encouraging enough to advance to the next trial phase or to market, N-of-1 testing is simply what it implies: each patient is assessed only on his or her individual response. N-of-1 isn't being advocated as a replacement for the more conventional between-group comparisons but as a useful complement where appropriate. And it can be performed with the same methodological rigors as applied to between-group studies, such as so-called "blinded" usage and assessment so neither the person who administers the drug nor the person judging its effects knows which dose or placebo was used on a given patient.

Some could argue N-of-1 is already the standard method for testing drugs for rare diseases (because much fewer patients exist) as well as prescriptions for compassionate and off-label indications. N-of-1 is likely to become more widespread as research involving precision medicine grows, in response to the recognition that we're all at least a little different from each other genetically, environmentally, and microbiologically, and these differences may influence how our individual diseases develop and respond to treatment. Certainly, there are circumstances in which N-of-1 is neither practical nor appropriate, such as public health trials to test vaccines and other preventive health measures for entire populations. But for the growing number of illnesses that delineate what's different between patients (or their individual tumors or pathogens) rather than what's shared by them, N-of-1 trials may provide major improvements in efficacy and safety at a societal cost that's reasonable.

The N-of-1 trial concept itself isn't new, even with respect to laboratory animals. It was popularized in the mid-20th century, under a different name, as the preferred experimental design of B.F. Skinner and other behavioral scientists who studied operant conditioning by repeatedly exposing animal subjects to various stimuli to see how they responded.[19] Each animal was studied individually and intensively over time, under what is known as single-case research design (i.e., N-of-1). Single-case research design was so closely identified with operant conditioning that the two were thought inseparable and not applicable to other lines of investigation. But single-case research design continues to be used in many scientific endeavors, applicable to any investigation in which interventions are repeated in the same experimental subject (i.e., the subject serves as its own control), and the subject is evaluated repeatedly over time.[20]

There is also an established version of sorts for N-of-1 animal trials in the field of regulatory testing. Acute oral toxicity determinations previously relied on $LD_{50}$ assays, the purpose of which was to identify what dose of the chemical (also known as the test article) of interest given to a group of animals killed 50% of them within 24 h after dosing. Multiple groups of animals were needed to estimate the $LD_{50}$, with deaths occurring in any groups that were close to the final answer. In addition, many other animals became sick and suffered whether they died or not. In the 1990s, the International Conference on Harmonisation of Technical Requirements for Pharmaceuticals for Human Use (known as "ICH") established that the $LD_{50}$ assay was no longer required by the regulatory agencies of the signatory countries. Soon thereafter, the FDA followed up with a new guidance that discouraged the use of the $LD_{50}$ and similar assays, sparing lots of animals from pain and death.[21] Instead of the $LD_{50}$ assay, other methods were promulgated that required smaller numbers of animals per group and fewer animals overall. One of these, the "up-and-down" assay, mirrors the N-of-1 approach because a single animal is given a specific dose and monitored for its response before a second animal is given a different dose based on the reaction of the first animal. If necessary, a third animal is given a dose that differs from the first two, monitored for its response, and so on. Adoption of the up-and-down assay has been credited for reducing animal use in acute toxicity testing in some situations by up to 80%[22] even though replacement of the earlier $LD_{50}$ assay was slow.[23]

So if there are scientifically established precedents for using N-of-1 in patient efficacy trials, animal behavioral research, and regulatory pre-clinical toxicology, can the N-of-1 approach be expanded to other animal-based research? Of course it can, especially in situations where animals or test article is very expensive, or where there may be a wider variety of responses within groups than between groups. Applicable situations for N-of-1 that come to mind include scarce lines of mice that are difficult to breed so few pups are available for an experiment (i.e., the rodent equivalent of a rare disease), or protocols that consume minute quantities of test article that are very expensive to produce: administering precious drug in smaller aliquots to fewer animals could possibly generate the same scientific conclusions. And N-of-1 (or N-of-a-few) studies are commonly performed when few clues are initially available to indicate what an effective dose may be so that the larger, "real" experiment is designed with appropriate forethought. Such pilot studies are rarely published but already serve a valuable role in avoiding wasting animals, test drug, and time. In all of these cases, one could employ an N-of-1 protocol design by enrolling individual animals on a sequential basis in experiments to reduce animal numbers and perhaps arrive at more precise conclusions faster.

Finally, let's return to Russell and Burch and all of their 3Rs. Replacement is often invoked these days whenever lower order animals are used instead of higher order ones for research or testing. This is most commonly applied to using laboratory invertebrates such as fruit flies and nematodes instead of mammals but has occasionally been (ab)used when NHPs are replaced with rodents. However, Replacement was originally defined as "any scientific method employing non-sentient material which may in the history of experimentation replace methods which use conscious living vertebrates."[24] So any alternative that still uses vertebrates of any kind[25] but in smaller numbers should rightfully be categorized as Reduction irrespective of phylogenetic difference. What if other vertebrates, namely veterinary patients, were used instead of laboratory animals to test new medicines designed for human patients with the same condition? Rather than induce an illness or injury in an initially normal animal to model the same condition in man, what if one could utilize pets that unintentionally were born with or acquired the medical problem of interest in human patients, and have the new drug, vaccine, or device given to them? That could legitimately qualify as Reduction if fewer animals were ultimately used. Another justification for such an approach is that if the test drug works in the animal trial, then the pet benefits as well.

Arguments against this alternative include the likely confounding variation between animal subjects one would have to tolerate in a given trial, versus the more intentionally homogeneous population of laboratory animals used in experiments. In addition, there's the ethical burden placed on pet owners to decide if the risk to the animal is worth the potential reward, as well as the related ethical burden of the trial sponsor to try to make owners grasp that the likelihood of their pets benefitting medically from this trial is slim. That's because a range of doses are usually involved in the trial and some, if not most, of these doses may not be efficacious and could be toxic. However, these latter issues are identical to the ones inherent in human clinical trials and are the reason that informed consent and institutional review boards were established. So there exist time-honored precedents and tools available for such pet-based trials. At least one firm was launched precisely for enrolling pets in pre-clinical testing protocols in advance of *human* drug trials.[26] If it and others succeed in using fewer animals than historically needed for pre-clinical drug evaluations, then we can celebrate the arrival of yet another Reduction option.

# Chapter 17

# When Humane Endpoints Aren't

Over 10 years ago I attended a national IACUC conference where the featured guest speaker was a prominent scientist who was studying the body's response to a severed spinal cord in laboratory mice. The goal of the research was to understand what happens at the cellular level and in what sequence after acute spinal cord injury so that similarly afflicted trauma victims may have a better chance of regaining function distal to the lesion. The talk was scientifically elegant and quite detailed, showing how various genes in various cell types at various distances from the spinal cord damage were activated or suppressed, and the downstream effects of these events. But in all the experiments presented the injured mice were allowed to decline to a state of paralysis followed by a moribund state, at which time they were euthanized. This made no sense to me. If you want to know what's going on in the spinal column, especially at the molecular level, during post-injury inflammation and neuronal death, and you already know that the lesion you are creating will result in immobility and eventual inability to reach food and water, why allow the animals to progress to such a terrible state? And what value are the data taken at that very late stage when the condition of the animal and its cells are more affected by hypothermia, dehydration, and inanition than any spinal cord injury?

Those were questions I posed in the course of my presentation to conference attendees the next day. But I already knew the answer. This animal model had been around for a long time, used by many laboratories, and relied on the response of the entire animal as a valid data point. So it was mostly because of convention and precedence that mice were allowed to decline and suffer further until they died. But if, as a scientist, you're interested in biological phenomena of a microscopic instead of a macroscopic dimension, why not euthanize these mice earlier before they become so weak? Or at a minimum, why not provide those mice supportive care in the form of parenteral fluids, nutrition, and an external heat source so they might live longer? If you are studying phenomena that occur during later stages of the body's response to spinal cord injury, it behooves your research to follow these mice for as long as possible. Had I been on the IACUC at that institution, I would have raised these points in reviewing that protocol or hopefully learned of extenuating circumstances that may have addressed my concerns. Sometimes there are sound reasons for allowing animals' health to decline, sometimes there are not. Sometimes those reasons are obvious as well as justified, sometimes they're not.

Notes in the Category of C. https://doi.org/10.1016/B978-0-12-805070-5.00017-5

In this chapter, we take head on the matter of intentionally harming animals for the sake of knowledge—for discovery, for testing, for teaching others. It's the central ethical issue of using laboratory animals. If there's no or minimal harm involved, there are likely to be no or minimal objections by most persons. But even though that's the reality for the majority of laboratory animals, if only one or a few are knowingly subjected to pain or distress, and although that is alleviated as best as we can, the fundamental gulf remains between persons who believe this use of animals is justified and persons who don't. Hence, it all comes down to what are known as endpoints, i.e., the threshold or degree of pain or distress that we as a society find acceptable (literally, the point in a given experiment at which the live animal is no longer of scientific value). At stake are millions of animals versus billions of persons who may or may not benefit from the results these laboratory animals may yield while some of these animals and many of those persons suffer. The spectrum of acceptability extends from outright abolition to giving scientists carte blanche (which, by the way, is not even considered an option today). We try to address this by invoking the 3Rs. While one (Replacement) avoids animals altogether, the other two (Reduction and Refinement) are attempts at compromise, imposing trade-offs that often fail to satisfy.[1] Since I addressed Reduction in Chapter 16, the focus of this chapter is on Refinement, the experience of the individual animal and how that experience could be less harmful (Replacement will be covered in Chapter 19).

Let's start by reviewing pertinent federal laws and regulations. Both the 1985 AWA amendments and the Health Research Extension Act of 1985 and their regulations compel scientists and IACUCs to employ the 3Rs when composing and reviewing protocols for approval.[2] However, the AWA also states that "Nothing in this chapter... shall be construed as authorizing the Secretary (of Agriculture) to promulgate rules, regulations, or orders with regard to the design, outlines, or guidelines of actual research or experimentation by a research facility as determined by such research facility." The accompanying Animal Welfare Regulations state that "Except as specifically authorized by law or these regulations, nothing... shall be deemed to permit the Committee or IACUC to prescribe methods or set standards for the design, performance, or conduct of actual research or experimentation by a research facility.[3] Similarly, the Health Research Extension Act of 1985 states that "Such guidelines shall not be construed to prescribe methods of research."[4]

When federally mandated IACUCs arose during the mid-1980s, scientists were understandably concerned about interference by these committees on experimental design and methods. It was commonly held that because IACUCs were unfamiliar with the nuances and current thinking about every animal model, the chances for damaging science, no matter how benign the intent, were much greater than actually improving an experiment. Furthermore, protocols were based on established scientific convention and years of dedication by experts. In addition, most proposed experiments had already undergone peer review, and animal-based assays were dictated by regulatory agencies such as the FDA or

EPA. So it was thought quite appropriate that IACUCs should be instructed not to "prescribe" experimental details in the course of their protocol reviews. On the other hand, how were IACUCs supposed to apply the 3Rs without considering if a reasonable number of animals are to be used, how pain or distress may be avoided or alleviated, or if animals were justified in the first place? These two sets of directives to IACUCs were in conflict the moment they were established. The real question was not if that conflict could be balanced but rather how long before it would be corrupted or ignored? Fast forward to 2004 when I co-moderated a workshop at a national IACUC conference on ways to evaluate scientific merit in animal protocols. In that session, I was pleased to hear from most of the audience that scientific value and study design were routinely addressed by IACUCs in their protocol reviews and considered fair game for revision. Contrast that with the 1985 standards that were launched with an expectation that scientific validity was not to be even mentioned during the protocol review process; only those components of a protocol that dealt with immediate and direct impacts on the animal subject were reviewable. As cumbersome and illogical as this divide was, because animal experimentation and animal welfare are so intertwined, it didn't take long (in regulatory compliance time) for it to be breached.

Other changes were occurring at the same time with regard to severe endpoints in the realm of animal safety testing. For decades, the dose of a compound that would kill 50% of the animals given that dose within a specified period of time (popularized as $LD_{50}$) was a required outcome for toxicity and safety tests intended for regulatory review. This standard was Refined in the 1990s by international agreement so alternatives to the $LD_{50}$ were now acceptable that either didn't kill as many animals or allowed laboratories to euthanize moribund animals that were likely to succumb anyway.[5] Over the years since, a consensus has developed that death as an endpoint comes with a very high bar for IACUC approval. It's now much rarer that animals are required to die on their own instead of euthanizing these animals when it becomes obvious they are dying.[6] But this positive change has resulted in an ethical stasis of sorts. Applying the label of "humane endpoint" to moribund euthanasia has become common in IACUC parlance even though that outcome isn't much more humane than letting an animal die. By the time an animal is moribund, think about the likely pain or distress that animal has experienced prior to reaching that state. It probably hasn't eaten or drank for hours or days, it can no longer maintain core body temperature so it's cold (unless it's febrile, which won't last much longer if it can't consume more calories to generate heat), and may be immobile because moving is too painful. Some IACUCs even apply Category D to these endpoints because there is an intervention prior to death that provides (terminal) anesthesia or analgesia.[7] By accepting this as "humane" instead of "only slightly better," I fear we've become complacent and won't explore alternative Refinements in protocol methods and endpoints as rigorously as we should.

Lots of opportunities for better Refinement remain unrealized if not unappreciated, partly in deference to scientists' legitimate claims of superior knowledge about their research fields and partly due to IACUCs' acceptance of preemptive euthanasia of moribund animals as a humane endpoint. What further endpoint Refinement could be gained? Take infectious disease models, for instance. Why not analyze body fluids, biopsies, or post-mortem tissue samples prior to severe illness in order to determine the number of pathogens or immunocytes or concentration of antibodies, toxins, or other aqueous molecules per unit volume or mass? If something is truly or even partially anti-infective, we already know that numbers of surviving microorganisms or their products will decline over the course of treatment. So why maintain inoculated animals any longer than necessary, especially positive control animals that will be inoculated but not receive any drug, in order to establish anti-infective efficacy after the pharmacokinetics and dosing regimens are established? If one is concerned that the batch of bugs inoculated into animals in a particular experiment or assay is less potent than required, certainly there must be surrogate indicators of pathogenicity that occur earlier or are confirmable in vitro or through genomics that correlate closely with severe pathogenicity in vivo.

The early days of the AIDS crisis offers a parallel example. For human clinical trials involving new AIDS drugs and in light of growing numbers of dying patients who had no hope at that time, FDA replaced its conventional endpoint of length of survival with AIDS (i.e., time of progression to death) with surrogate markers that were just as reliable in predicting drug efficacy but changed much earlier than an irreversible decline in health. The two most popular surrogate markers were the number of CD4+ T cells and HIV-1 RNA levels in the blood, tracked during a trial to see if CD4+ T cells didn't decline as quickly or at all and if HIV-1 RNA was reduced. While not perfect, the FDA considered these outcomes adequate for market approval in light of the national health crisis at that time.[8] If such applications are based on sound science and medically acceptable for human diseases, what's wrong with considering similar applications for animal models of those same diseases, even if animal research isn't considered as dire and we're given more leeway to push animals closer to the edge?

Moving on to cancer: blood samples, whole or partial body imaging, biopsies, or post-mortem tissue samples could be collected prior to the onset of severe illness to determine the number and location of tumor cells, followed by in vitro assays for changes in tumor or immune cell behavior or markers. The same logic presented above for infectious disease applies to oncology, i.e., if a trial product reduces tumor mass, metastasis, etc., that response alone could be enough to consider it truly anti-neoplastic. Animal models of any other type of disease or injury can be approached in the same manner, whether it's rheumatology,[9] nephrology,[10] or others.[11]

What about non-specific indicators of pain or distress not tied to specific models? On a practical level, simple visual cues have been described that indicate when animals are experiencing pain or distress.[12] These could be considered as either endpoints themselves or important initial warnings that animals expressing these cues should either be monitored more frequently, administered more or different analgesics, or provided other supportive care. Simple invertebrates may suggest another approach that involves physiology instead of behavior. When environmental conditions become incompatible with normal biological functions, such as intense heat, cold, or drought, the larvae of *Caenorhabditis elegans* and other nematodes convert to what's known as dauer formation or dauer arrest to shut down their metabolism and growth until conditions improve. Cellular receptors for detecting the onset of conditions incompatible with normal life are believed to lie in the gut of these worms and may have analogs in *Drosophila melanogaster* that detect the onset of harsh conditions and initiate a protective response.[13] If these genes or their analogs are conserved across taxa to mammals, could the expression of these genes or the proteins they produce serve as an indicator of significant pain or distress? I once heard a kidney scientist describe a genetically engineered mouse that would excrete green fluorescent protein into the urine under specific metabolic circumstances, and assigned the clever nickname, "G-F-Pee." Imagine how convenient it would be if a laboratory animal was engineered to excrete a dauer-like compound in its urine that was linked to a different color or other easily detected marker whenever the brain perceived severe pain or when glucocorticoid levels in the blood rose in response to chronically high levels of physiological distress. Because animals are so good at hiding pain or distress to avoid tipping off predators, an obvious indicator of their bad state could be helpful as a warning signal or endpoint before the animals' clinical status has noticeably changed.

If early intervention isn't an option, are there alternative opportunities to reduce pain and suffering as they become more severe? One approach hinted at above involves supportive care, in which universal treatment modalities are employed alongside drugs that target the disease or injury of concern. Those modalities include parenteral fluids and nutrients, analgesics, heat, quiet, darkness, oxygen, and other components intended to ameliorate the consequences rather than the cause of the illness. An intriguing extension of this arises from reports of empathy demonstrated between paired rodent conspecifics, one in pain and the other not. The mouse or rat not subjected to a painful procedure responds as if it were in pain, too.[14] Could it follow that the animal in pain could derive comfort from the presence of a compatible cage mate?

If you think I'm being impractical, don't confuse supportive care with intensive care. Many elements of the former are included in the latter but I'm not advocating routine use of a saline intravenous (IV) drip or oxygen tent for animals if they won't tolerate it or if doing so puts the animal or animal care provider at risk. But many types of supportive care are simple, safe, and cheap. Common arguments against employing supportive care for laboratory animals in severe pain or distress include interference with scientific objectives of the study, creating too much variation that's not easily standardized between animals or treatment groups, and destroying continuity with historical data for the same model that did not employ supportive care in the past. My riposte to these objections presumes continued progress in science and medicine.[15] If one is modeling a disease that occurs in human patients, wouldn't you want to recreate as much of the patient experience as possible to get as representative an outcome as possible? And wouldn't you want to sustain that animal as long as possible with enhanced (supportive) care to extract more data over a longer period of time if earlier euthanasia is not an option? Furthermore, the difficulty in standardizing supportive care between animal subjects mirrors the same reality amongst patients, yet we still seem to generate dependable scientific conclusions and regulatory decisions from clinical trials and retrospective studies of sick people. Finally, if you never included supportive care in past versions of an animal model, does that mean you've never modified that model in any other regard, either? Upgrades and enhancements, whether they're of a scientific or ethical origin, should be given serious consideration for animal-based experimentation and testing involving likely pain or distress.

Beyond early warnings of generalized pain or distress, can we adopt equations and tools from the field of critical care medicine to predict more precisely the fate of a laboratory animal in serious medical decline as an eventual outcome in a protocol? Common examples in the human medical sphere include Acute Physiologic and Chronic Health Evaluation (APACHE), Simplified Acute Physiologic Score, and Mortality Prediction Model. An example of their predictive accuracy is that APACHE version III scores predicted within 24 h after admission to an ICU the likelihood of hospital death within 3% of actual outcomes for 95% of over 17,000 patients admitted to 40 US hospitals.[16] APACHE is now on version IV.[17] It is available on a website that can be accessed via smart phone for quick and convenient calculation of Estimated Mortality Rate and Estimated Length of Stay.[18] Similar clinical scoring systems have been developed for dogs and cats in advanced states of disease and injury.[19] Let's turn these tools around 180 degrees in a sense and explore their power for predicting mortality in laboratory animals expected to die eventually from a protocol-induced insult. In other words, could cage-side observations, vital signs, and clinical laboratory values be compiled for a predictive algorithm so that an animal's expected decline is assigned a score earlier in order to intervene sooner with euthanasia? Returning to infectious disease, this combination of inputs is already described for common animal models of sepsis, such as injecting fecal contents into the peritoneal cavity or ligating and then puncturing the cecum. In either case, mice will die from organ failure unless there is a successful medical intervention or they are euthanized first. The same holds for total body irradiation at high doses that leads to sepsis and death in the absence of treatment, such as a bone marrow transplant. Reviews of clinical prognostic markers for potentially lethal organ failure associated with sepsis should be scrutinized to see how much earlier, and thus more benign, endpoints may be applied more aggressively to animal models.[20]

While supportive care may extend the life of a patient while also reducing their discomfort, a second related approach to reducing severe pain or distress for those dying is hospice care. Concepts such as end-of-life and quality-of-life have been articulated for laboratory NHPs never expected to recover and likely to die soon, and seem worthy of consideration for all sentient animal subjects.[21] Guidelines for veterinary hospice care are widespread[22] and the extensive human medical literature offers much to digest in this respect.[23]

How else may be protocol endpoint review and approval by IACUCs be made more exacting, perhaps requiring additional yet justified explanations by principal investigators (PIs), so that moribund euthanasia becomes passé as a "humane endpoint"? One detail that deserves attention is the inadequacy of non-affiliated (community) and non-technical IACUC members to fulfill their responsibilities when it comes to understanding and passing judgment on scientific details of a protocol. These well-meaning persons perform an impossible task, if society's expectation is for them to provide an effective counter-weight to insiders' interests (laboratory animal veterinarians included) and ambitions. On the other hand, if they aren't expected to do that, then they're just a rubber stamp for committee decisions. The reality is that most of these members aren't scientists or veterinarians themselves so they get quickly lost in the jargon and depend on others for interpretation. Or they are intimidated by the knowledge and credentials amassed around the table and don't want to appear too ignorant. So they don't participate to the degree the public or regulators may expect, or they're too deferential even if they open their mouths.

Conversely, it's been my experience and observations that scientists and laboratory animal veterinarians are quite comfortable and willing, as IACUC members, to challenge elements of a protocol that don't appear at first glance to make sense. They are often quick to point out protocol details that seem weak and could be improved with regard to both the 3Rs and sound experimental design principles. But those familiar with the care and use of laboratory animals are discouraged, if not expressly prohibited, from serving as non-affiliated members on another IACUC.[24] This is presumably out of fear that we'd be too lenient under the guise of professional courtesy and cut someone an unwarranted break. But I've seen just the opposite behavior time and time again.

Since we're stuck with the explicit ineligibility of non-affiliated members who are more knowledgeable and therefore more likely to provide constructive input in protocol endpoint deliberations, what can be done to improve the effectiveness of conventional lay IACUC members? For one, don't stop at the usual on-boarding and training of lay IACUC members beyond a recitation of the regulations, giving them a copy of the AWA and the Guide, enrolling them in introductory courses on laboratory animal care and use provided online such as the AALAS Learning Library and the Collaborative Institutional Training Initiatives Program. Where are the (advanced) lay IACUC courses to improve familiarity with life sciences, statistics, bioethics, and animal research? Through an institution's extension school and online learning platforms, completing suitably designed courses to elevate lay IACUC members to a higher level of understanding should be required, or at the very least, rewarded. I know these folks are volunteers, and good lay IACUC members are hard to find. But we still need to build their competencies in order for them to fulfill their responsibilities as expected by society and described by regulators. A great place to start is an overview of what is considered acceptable experimental design. In addition to enhancing the lay member's contribution, such knowledge in robust IACUC deliberations could help correct the current crisis in data reproducibility, in which poor experimental design and sloppy conduct are considered major contributors.[25] Imagine, as a result, a non-affiliated or non-technical IACUC member piping up to ask if experiments will be performed blinded, are there (suitable) positive and negative controls, are animals properly randomized to the various experimental groups, and are reagents validated prior to their use in animal subjects? If the answer is "no" to any of these questions, then someone on the IACUC should ask why not?

In addition to advanced learning opportunities, consider employing a strategy that's becoming popular to ensure patients understand their drug prescriptions and discharge instructions. Known as "teach-back," this is where the person receiving important information (a patient or family member) repeats it to the information provider (a physician, nurse, pharmacist) as if the former were teaching the latter.[26] The objective is to show that the patient truly understands what he or she has been told. It forces medical professionals to avoid jargon, speak clearly and slowly, and encourage questions. If the patient can't "teach" the instructions given, then it's likely those instructions won't be followed. In part because of its demonstrated effectiveness in patient understanding, teach-back is getting attention for informed consent in human clinical trials, in which the patient or patient's guardian "teaches" someone on the clinical trial's team about possible risks as well as possible benefits for enrollees that are supposed to be spelled out clearly in the trial protocol and informed consent document.[27] The same could apply for IACUC protocols, in which the lay member "teaches" someone with more scientific literacy what the salient points are in the protocol and what is the risk of harm to the animal subjects versus the potential benefit of the proposed animal use. It would interesting to see how well this approach could work.

On a different tack, there's nothing in the rules that prohibits adding an IACUC member who is not affiliated with that committee's institution but has expertise in the care and use of laboratory animals, as long as someone else serves as the official non-affiliated member who has no prior record of animal research-related activities and thus remains chaste. If some IACUCs were willing to add the former to the latter, does that offer an opportunity to judge which membership composition is more rigorous in reviewing protocols (with "more rigorous" defined as more likely to suggest procedural and endpoint Refinements to the PI)? To answer that question, I propose conducting a social experiment using four groups of IACUCs. The first group would increase their respective memberships by adding an experienced laboratory animal veterinarian[28] who is not affiliated with the institution, while continuing to engage their official lay members the same way they always have. The second group would employ the teach-back method with their respective lay members to establish comprehension of protocol rationales and hopefully encourage more dialogue from those lay members during protocol reviews. The third group would combine what the first two groups are doing, i.e., add an outsider laboratory animal veterinarian as well as employ the teach-back method for its lay members. The fourth group would serve as negative controls, with no membership additions and no changes to how lay members are trained or engaged in committee deliberations.

Each of the four groups would consist of 25 IACUCs, to be recruited from USDA institutional registration rolls and then confirmed to have an Assurance on file with OLAW and be AAALAC-accredited. That way, each IACUC in the experiment would be familiar with both AWA and US Public Health Service (PHS) Policy details and subjected to regular external reviews for compliance to the same regulations and accreditation standards. There are close to 1000 USDA-registered research facilities,[29] over 1200 US institutions with a PHS-approved Animal Welfare Assurance on file,[30] and over 660 AAALAC-accredited units in the United States.[31] So there should be plenty of candidate IACUCs that could participate in this experiment. Social scientists could help normalize each of the four groups for institutional size and purpose (i.e., for-profit product R&D, including contract research organizations, animal breeders, government research laboratories, universities, and academic medical centers), species mix, and size of animal census. Enrollment and assignment to groups could be done on a rolling basis rather than starting everyone at the same time; each IACUC would have to keep their assigned protocols and deliberations secret so subsequent enrolled IACUCs wouldn't be influenced. Social scientists could also advise on instructions and monitoring for compliance to the experimental plan, as well as devising pertinent metrics (endpoints!). Perhaps surveys of all IACUC members before project initiation and after project completion could be informative, showing how perceptions of committee effectiveness may or may not have changed. A less subjective and likely more relevant metric would be to track the number and types of changes made to each protocol or amendment from the time it's submitted by the PI to when it's officially approved by the IACUC.

My hypothesis to be disproved is that protocol endpoints will be more Refined yet still scientifically acceptable (just as a critical an outcome) when either additional experts participate or teach-back is used on lay members. It's a given that many elements of this project need to be resolved before it should occur. Who will gather and analyze the data? How will it be shared with other participants and the public? How will PI and IACUC confidentiality be protected? What effect, if any, will there be on protocol reviews when an IACUC knows that reviews will be scrutinized by outsiders? Will an IACUC's assigned group influence how that IACUC will deliberate? How much lead time should be provided to an IACUC that would have a different (expanded) membership or different (teach-back) process than it's used to, before its protocol reviews enter the project data stream? How long should each IACUC participate? What if an IACUC and its PIs had already achieved a state of thorough protocol review with truly humane endpoints, so very few new Refinements occur or are necessary after being assigned to a group—would that skew the composite data and erroneously disprove the hypothesis? And finally, if the results indicate one approach may be better than others, would the regulators insist every IACUC adopt that approach even if it may not be appropriate for every institution? Despite all these questions, I still think it's worth trying. One should not infer from my proposal that every protocol has experimental design or endpoint liabilities; most of them, if fact, don't. For those protocols that do and even if they all conform to appropriate scientific standards and no further Refinements are possible, what's the harm in verifying this via a more informed and engaged committee review?

Another approach to consider how endpoints may be come more humane is to convene a bioethics conference to discuss the following hypothetical: what would happen if moribund euthanasia and death without intervention were no longer allowed as protocol endpoints? Which animal models and therapeutic fields would be impacted the most? What additional knowledge would be needed to fill the resultant gap and get back on track with respect to medical progress under those restrictions? How hard would it be to gain that additional knowledge, how long could it take, and how much could it cost? Much about these questions is unknown, but by merely asking them in a public forum we're challenged in a good way to come up with better answers than merely pointing to the pain and suffering of patients and why that justifies what we do.

Taking that bioethics conversation to an even more restrictive plane, what scientific and medical progress is at risk if animal research and testing could involve no detectable pain or distress? This line of thought, articulated by some animal protectionists, prohibits painful procedures on an animal subject unless these procedures likely (or possibly?) improve the health of that animal.[32] While well reasoned if not reasonable, adopting this tack would essentially convert IACUCs to institutional review boards, heightening the animal's stake in deliberations by a local research oversight committee. One consequence of this approach could mirror the basis for clinical research protocols in that animal subjects, like human or veterinary patients, would not involve a no-treatment (negative control) group whenever a new medical product is being evaluated for efficacy. Instead, the control group would have to be provided the current standard of medical treatment and care, as the basis for comparison with the experimental treatment groups of interest. Any positive outcomes in the latter groups would be evaluated for statistical significance against these animals being administered a drug, vaccine, or device already approved or used for that condition.

Setting aside the obvious and substantial damage to biomedical research as we know it today, adopting this stance could possibly incorporate informed consent, a fundamental component of clinical research intended to ensure the research subject is sufficiently knowledgeable about the risks as well as the potential benefits of enrolling in a clinical trial. The human clinical research field offers an alternative for patients who are not able to provide their consent, such as infants, unconscious adults, and dementia cases; persons considered acceptable surrogates to consent on behalf of those deemed unable to understand are usually family members, and less often, court-appointed guardians or others, usually determined by state laws.[33] For pets enrolled in new veterinary drug trials, the pets' owners are accountable for informed consent.[34] In an animal research protocol, who would be responsible for consenting on behalf of laboratory animals? Under what circumstances could an animal protectionist be granted advocate status for an animal? Should that guardian be a veterinarian who would likely understand better the potential medical benefits and risks involved? Exploring this approach and its ramifications in a respectful and constructive forum may be enlightening even though consensus from all parties is highly unlikely, to say the least.

Finally, there's also a possible opposite outcome to all of the above as a result of more meticulous IACUC deliberations over acceptable endpoints. Some animal models of medical importance that previously wouldn't be allowed because their conventional endpoints were deemed unacceptably severe could be reactivated by an IACUC perhaps more energized in its deliberations and more confident of its judgments. Take for example, the study of burn wounds and their sequelae for the purpose of developing better treatments. It's been my observation that animal models of second-degree burns were permitted by IACUCs with certain caveats, whereas third-degree burn models weren't due to the latter's presumed greater degree of pain in animals. Yet severe burns certainly occur, especially among children; one study showed that one in five burn patients between the ages of 2 and 13 years, with at least 30% of their body involved, were likely to experience multi-organ failure.[35] Given these realities, why not model more extensive burns in animals but incorporate earlier endpoints plus appropriate supportive care?

Chapter 18

# The Hardest Field

*"How do you balance the needs of the animal against the needs of the investigator?"*

This question was posed to me cageside by an animal care technician one day while both of us were looking at a very sick mouse. I was called down to the vivarium for a consult because the animal could not move its hind legs and was in worse shape than earlier that week. This particular mouse had been genetically engineered to develop amyotrophic lateral sclerosis (ALS, commonly known as Lou Gehrig's disease), for which there is no effective treatment for human patients.[1] At the time of the examination, the mouse was presenting with clinical signs characteristic of advanced ALS, such as obvious weight loss and weakness so severe that the animal could only get to food and gelled water on the floor of the cage by dragging itself around by its forelegs. The protocol to which this mouse was assigned allowed for afflicted animals to progress (descend?) to such a state while they were given experimental therapies to see if the ravages of this murine version of ALS could be reversed or just slowed down. It was argued in the protocol that if these mice were euthanized at an earlier stage of the disease, before things got this bad, the actual efficacy of the therapy would be harder to detect and animals would have been wasted in a scientifically weaker experiment. In permitting this to occur, the IACUC imposed limits so that any animal approximating this particular mouse's condition was to be euthanized as soon as possible in order to spare it from further distress. Because veterinarians attending to laboratory animals, at least in the United States, have the legal authority to intervene on any animal's behalf at any time, consults are common when protocols involve unalleviated pain or distress such as this one. The decision I had to make was had this mouse crossed the threshold requiring immediate euthanasia or should it be allowed to survive a little longer so more and perhaps critical research data could be obtained per the researcher's intent?

Newly minted veterinarians are administered the Veterinarian's Oath during their commencement ceremony. When I graduated in 1982 from Washington State University, the Oath at that time read as follows: "Being admitted to the profession of veterinary medicine, I solemnly swear to use my scientific knowledge and skills for the benefit of society through the protection of animal health, the relief of animal suffering, the conservation of livestock resources, the promotion of public health, and the advancement of medical knowledge. I will practice my profession conscientiously, with dignity, and in keeping with the principles of veterinary medical ethics.

Notes in the Category of C. https://doi.org/10.1016/B978-0-12-805070-5.00018-7

I accept as a lifelong obligation the continual improvement of my professional knowledge and competence."[2] Each societal benefit listed in the Oath spoke to a different field of the profession. For example, "promotion of public health" pertained in part to making sure foods of animal origin, such as meat, eggs, and milk, are safe to consume while "protection of animal health" encompassed development and deployment of vaccines as a reliable preventive health practice. Two of the other benefits listed i.e., "relief of animal suffering" and "advancement of medical knowledge," point to the intersection in which laboratory animal medicine exists.

I try to take the commitment in the Oath to relieve animal suffering to heart every day. But how can it always square with practicing in the only field of the profession in which some animals are intentionally subjected to illness or injury, and ameliorative treatment is intentionally withheld from some of these, all in pursuit of better treatments or cures for human and animal patients? At what point should an animal be euthanized or withdrawn from a study for humane reasons, and if so, at what intellectual and moral cost to human knowledge and medicine? The rub, as they say, that comes with working in this intersection arises from the uncertainty of research. Sometimes science moves forward smoothly, sometimes haltingly, and sometimes it hits dead ends and blind alleys. And you would not know the outcome until after the experiment is concluded. That is why the experiment has to be actually performed rather than merely conceptualized. Even then, the results may not be clear but argue for another try, perhaps with a slight modification but again with no guarantee of enlightenment. Of course, the stakes are higher when animal pain or suffering is involved, especially when that pain or suffering is induced and even more especially when it can not be alleviated for scientific reasons. That is one reason every protocol reviewed by an IACUC has to justify the number of animals requested, including their assigned pain or distress groups, as well as attest that acceptable replacements replacements do not exist and that the research proposed does not unnecessarily duplicate other experiments.

Most, if not all, laboratory animal practitioners I know have negotiated some sort of personal reconciliation between these two opposing responsibilities, i.e., assisting in the imposition of pain or distress on animals versus minimizing that pain or distress when appropriate. It's not something we talk about much, if at all, but it is understood that a never-ending ethical balancing act comes with the territory. I suspect that others' reconciliations, like mine, are based on two intertwined convictions. First, that the potential benefit to patients, their families, and society is so compelling that it is worth the immediate and occasionally severe animal welfare costs. That's acceptable as long as the second belief is also legitimate, that we can help make a difference to ensure that each animal subject is used to its fullest value so that it is not wasted and that experiments can and will be refined so that the animal subject is spared possibly worse experiences. In other words, it's our job to give the animal as soft a landing as possible without compromising the scientific aims of the study and to advocate on behalf of that animal when differences in opinion arise. Michael Fox (the veterinarian, not the actor) framed this intermediary role well when he described veterinarians "as arbiters between society and animals" and the veterinarian "as an interlocutor between human interests, animal interests, and the greater good,"[3] although I suspect he and I would disagree on what qualifies as the greater good.

What if these two moral convictions don't hold? Let us start with the requisite faith that the ultimate potential benefit is worth the cost. If you don't believe that, you shouldn't be in the game. When I speak to veterinary students interested in laboratory animal medicine, I raise this moral dilemma as a central piece of the practice and advise that the field is not for everyone. I tell students that if you get into it but your head and heart can't find a middle ground, then you should get out. The good news is that the vast majority of laboratory animals are not subjected to intentional and unalleviated pain or distress; most animals in a vivarium on any given day are either breeding (especially mice and zebrafish), or awaiting a procedure that is usually minimally invasive (such as a blood draw) or a humane death. Nevertheless, it's the infrequent situation that tests one's moral convictions and resolve.

Conversely, it's even more important to leave the field if one's compassion wanes from boredom, burnout, or wasn't strong enough in the first place.[4] If you no longer feel at least a little uncomfortable when examining these animals or watching them being euthanized, please find another line of work pronto. To stay engaged is not only a disservice to the animals and investigators but taints the rest of us as well. The same perceived callousness could also be possible if your ethical prism was shaped by cultural values elsewhere in the world but does not match those of the contemporary American society.[5] Fortunately, such loss or absence of compassion is extremely rare and eventually rectified by external pressures if that veterinarian does not self-correct or switch careers.

How about the second conviction that we can and do make a difference? Providing investigators high-quality animals free from natural diseases and maintaining them with good food, drinking water, and comfortable housing conditions is a start. These parameters comprised the initial stimulus that launched this specialty back when animal health status, nutrition, and housing were not standardized.[6] And even though ethical responsibilities have been integral to the Veterinarian's Oath from the beginning, veterinarians were often the sole advocate for laboratory animals if disagreements arose over what the experiment needed, how much money was available, what was practical, and how animal welfare would be impacted. Sometimes veterinarians were a minority of one in these discussions and consequently outvoted or overruled. It was not until the mid-1980s that more backup arrived after IACUC oversight became required, sometimes boosted by the participation of non-affiliated and non-scientist voting members. But what happens if your position or advice is rebuffed and without acceptable (to you) justification? Sometimes we just don't win the debate over a given protocol. Or the institution is unwilling or unable to upgrade its facilities per your recommendations, possibly risking regulatory non-compliance and losing AAALAC accreditation. Then it may become a professional and personal decision to stay or leave. How much opposition and adversity can you take, and for how long? Are your views reasonable? Have you offered practical solutions? What are your local job options, especially if your family is settled or other circumstances tie you to that community? The grapevine is quite active and institutions acquire reputations as either supportive of good laboratory animal care or not. New senior leadership at an institution can change things, sometimes for the better and sometimes for the worse. Therefore veterinarians need to pay attention to institutional mores and follow their internal moral compass if their voice is not being heard.

Even if one attains a professional accommodation in dealing with intentional and unalleviated pain or distress in animals, the eventual emotional costs can be substantial. One prominent and talented laboratory animal veterinarian I know left the field precisely because he could no longer stomach the large numbers of infant genetically engineered mice with unwanted genotypes that were euthanized in his institution.[7] Another very dedicated colleague got out after a few years because she could not tolerate how pre-clinical toxicity testing of candidate human drugs left too many animals in extremis without timely relief. The ability to cope or not with this emotional burden has been studied in a sociological context, most notably by Arnold Arluke at Northeastern University. He focused on the beliefs and feelings of animal care technicians who have to observe animals every day and are usually the ones who euthanize them, and on the attitudes of the investigators who use the animals in their research projects. The findings of Arluke and his colleagues highlight what they described as a "division of emotional labor" as a personal coping mechanism that is neither easily acquired nor maintained by some.[8] However, the laboratory animal veterinarian, who is squarely in the middle, is curiously absent from such investigations, even when such studies are published in leading veterinary journals.[9] I am not sure what, if any, novel insights could be generated from analyzing us, but our exclusion from such research is puzzling.

In closing, I'll return to the question at the beginning of this chapter posed to me while evaluating that poor mouse: "How do you balance the needs of the animal against the needs of the investigator?" My answer: "With great difficulty."

# Chapter 19

# Buggy Whips and Telegrams?

Every year PRIM&R convenes a large 2-day conference devoted to the oversight of the care and use of animals, encompassing IACUCs, federal agencies, and AAALAC. Over the past 10 or so years, these gatherings have focused more on *process*, such as how IACUCs may perform their responsibilities more effectively or efficiently, and how to interpret and abide by the latest utterances from regulators or accreditors, with less and less time dedicated to *purpose*, i.e., why are we (still) using animals, and is that use (still) justified? I've attended and spoken at many of these conferences but remember a recent one in particular. During a break between presentations, I was chatting in the front of the auditorium with a prominent veterinarian and long-time colleague and we began surmising how long will it be before laboratory animals are no longer needed? We agreed that 100 years from now is a sure bet, 50 years is likely, 25 years perhaps premature. I pointed out that we may be the only two in our field who were comfortable having such a conversation, to which my colleague responded with a wry smile. We both knew that while everyone in our field wholeheartedly agrees that animals shouldn't be used for research and testing if they're not necessary, it's considered heresy by some merely to acknowledge the eventual obsolescence of laboratory animal use as we know it today. One reason for these strong objections is that it's perceived as a current threat to one's career choice and job security, even though such an outcome is likely decades away. Another reason is that such a perspicuous admission gives animal rights activists more ammunition for their demands to abolish animals from laboratories without further delay.

Regardless of one's fears or hopes, there are accelerating trends and new technologies that will chip away at the traditional reliance on laboratory animals. This final chapter will review these trends and how they impact the immediate and not-so-distant future of laboratory animal use. Also to be considered is at what stage in that evolution are we currently? Have we crossed a tipping point or still immersed in business as usual, and does that matter? Just as intriguing, in what direction will societal attitudes change if the need for laboratory animals declines? How will a roughly 200-year run (1880–2080?), starting from the time of Pasteur and Koch to some inevitable finish, be viewed centuries from now?

Notes in the Category of C. https://doi.org/10.1016/B978-0-12-805070-5.00019-9

**189**

Let's start by looking at trends in the number of laboratory animals used in research, testing, and education. In the United States, that's not easy because the AWA explicitly excludes most rats and mice, and all birds and poikilotherms. As a result, these animals aren't included in the annual tabulations submitted by institutions registered with USDA.[1] But for species covered by the AWA and reported to USDA, the number of laboratory animals actually used is at its lowest level ever, from a peak of 1.9 million in the mid-1980s to slightly over 800,000 in 2016, the most recent US government fiscal year for which such numbers have been tabulated at the time of this writing.[2] This has been a gradual rather than precipitous decrease and the result of several factors. First, mice remain the most popular laboratory mammal because they're smaller and cheaper to maintain than most species regulated by the AWA and, up until now, have been the easiest to genetically engineer for increasingly sophisticated biomedical investigations. Consequently, mice have served in many regards as substitutes for larger, USDA-regulated mammals. So a decline in the usage of those animals is offset, in part, by switching to a more cost-effective and malleable species. Second, some research and testing on species covered by the AWA in the United States has shifted to developing countries that offer lower costs and less regulatory scrutiny or public transparency. Finally, some of the drop in laboratory animal numbers in the United States is simply because animal assays of yesteryear have been supplanted by alternatives that require no animals at all. A good example of this is the use of mostly rabbits up through the 1970's to produce polyclonal antibodies, which switched in the 1980s to using mice to produce monoclonal antibodies that were then replaced with in vitro cell cultures in the 1990s for creating the same types of reagents.

Can we get a more comprehensive snapshot of where laboratory animal numbers are going? Fortunately for this purpose, the UK's Home Office reports on the number of animals, including all vertebrates and cephalopods, used every year in research, testing, and education in Great Britain (including agriculture and wildlife, excluding research on new veterinary medicines). Data from 2015 indicate that 4.14 million procedures were completed on 4.07 million animals that were used for the first time. One-half of these procedures and animals involved actual use in research, etc., whereas the other half occurred to "create/breed genetically altered animals that were not used in further procedures."[3] Fig. 3 in the 2015 report, "Total experiments/procedures, 1945 to 2015," indicates that the actual animal use peaked in Great Britain in the 1970s and has since dropped approximately 60% by 2015, excluding genetically altered animals that were never used further. How to compare sustained decline in laboratory animal use in Great Britain to what may be happening in the United States and other wealthy nations with equally advanced biomedical research infrastructure? One approach is to normalize peak animal usage according to data from countries that report all species, then plot these numbers over time against corresponding changes in the number of USDA-regulated animals reported in the United States in order to deduce how total animal usage in the United States may be changing. This is reasonable because the ways in which laboratory animals are used in research, testing, and education in wealthy nations varies little, unlike the differences in the scale of animal use between these countries.[4] If the premise of this approach is acceptable, then its outcome shows how remarkably similar laboratory animal use trends are between the UK, US, and the Netherlands from 1960 to 2010, as depicted in the following graph.[5]

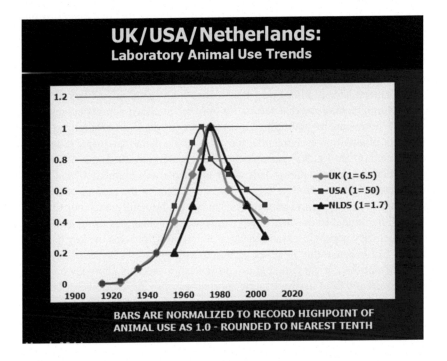

A finding from the 2015 Home Office report that's cited above illuminates a reality that's just as prevalent in the United States as in Great Britain but not as well enumerated here, i.e., the large number of genetically engineered mice bred but not used. This is an unfortunate byproduct of simple Mendelian genetics. The recessive genotype needed for a particular experiment may require several crosses over several generations of genetically engineered mice before a scientist has the composite genotype he or she is seeking for an experiment. If there's only one recessive genotype of scientific interest in a given litter, that usually means all the other newborns with unwanted genotypes aren't needed. So they are euthanized soon after their individual genotypes are determined, usually before they're weaned. To keep these mice around would quickly overwhelm the housing capacity of the facility and the resultant costs would be unaffordable. Performing euthanasia is never a pleasant task, and for a large research institution, dozens to hundreds of mice with unwanted genotypes are killed humanely every week. It would be great if someone developed some molecular biology razzle-dazzle that could terminate or prevent undesired mouse genotypes in utero so these embryos wouldn't go on to be born and then need to be euthanized 1–2 weeks later. Instead only, the remaining fetuses with the desired genotype would remain healthy for eventual birth and use. Of course, one would have to be careful with such technology so that it wouldn't be abused to select for or against specific genotypes in us (if it works on a mouse fetus, why not a human?). But current laws that ban human germline cloning and other molecular eugenic temptations could be augmented if necessary to avoid such Huxleyian abuses to our species.

There's even a better solution that is just beginning to emerge which should dramatically reduce the numbers of genetically undesirable mice that have to be euthanized shortly after being born. Known as "gene drive," this is a new gene-editing technology that selects for only the intended genotypes, resulting in 100% homogeneity amongst progeny as early as the $F_2$ generation.[6] It's getting a lot of attention as a potential means to tackle malaria, a major scourge of humankind. Over 400,000 persons died from malaria in 2015, out of 212 million malaria cases, and almost half of the entire planet's population was at risk of infection.[7] This is despite lots of effort and money devoted to developing new drugs that try to avoid parasite resistance and, hopefully, vaccines that will succeed in clinical trials. Genetically engineering mosquitoes using gene drive so neither they nor their larvae can carry malaria may provide the key to eradication. But because this involves tinkering with natural ecosystems that provide few options for containment or corrections, a cautious approach has been emphasized and field trials are being approached with lots of deliberation.[8]

In the vivarium, no such barriers exist to adopting gene drive for genetically engineered mice and other vertebrates. Imagine how laboratory animal numbers could drop if only genotypes of scientific interest were produced. Extrapolating from the Home Office's 2015 survey, one could anticipate a reduction in the number of mice by half(!) just from gene drive alone. The gene-editing technology on which the gene drive is based, CRISPR, promises efficient manipulation of genomes without all those mice wasted from Mendelian inefficiencies.[9] Other species besides mice are also attractive for CRISPR-enabled gene editing that could enable all sorts of novel animal models created with more biological precision than ever before when compared to older methods.[10] That in turn could result in an *increase* in the number of laboratory animals that aren't mice. Imagine rats engineered with better genetic lesions for human-like depression and bipolar disorder, or pigs with xenotolerant human-sized organs that wouldn't be rejected in many of the 117,000 patients awaiting transplants in the United States alone.[11] And that's just for starters.

However, it's likely that gene editing will complement, not replace, older genetic engineering technologies that continue to create valuable animal models that contribute to our knowledge of disease and medicine, such as a recently described transgenic monkey model for autism.[12] Another development that may increase laboratory animal numbers in the short run is the recent mandate from NIH to include female animals in experiments.[13] Historically, males have been more commonly used because females were retained for breeding and because, frankly, most scientists just didn't pay much attention to sex differences that could affect drug metabolism and other outcomes. But no matter how popular CRISPR becomes or how many female animals are added to experiments, any resultant increase in animal numbers will be counter-balanced by a net savings in animals from gene drive if it pans out as envisioned, thereby leading to further declines in laboratory animal populations. One can expect occasional excursions away from this trend, such as using more rather than fewer laboratory animals when emerging infectious diseases threaten to become pandemics, such as Ebola or Zika virus.[14] Or if someone develops better animal models for human dementias that are forecasted to overwhelm national health care budgets.[15] But these episodes will be temporary exceptions to a continuing net reduction in laboratory animal use.

On the medium-term horizon, say over the next 10–15 years, animals will be replaced in ever greater numbers by non-animal technologies. One much anticipated area is toxicity testing that will use faster, cheaper, and just as informative (if not more so) technologies that involve no live animal-based assays. In 2004, the US National Toxicology Program (NTP) published its vision of how high throughput robotic screening that relies on bench chemistry and cell culture could accelerate safety assessment of thousands of chemicals awaiting regulatory judgment.[16] That document and other initiatives generated a major collaboration between the FDA, EPA, and NIH (including NTP) known as Toxicology for the 21st Century (or "Tox21") with the intention of evaluating over 10,000 environmental chemicals in a comparatively rapid time frame with little need for new animal-generated data.[17] These developments have been mirrored by government agencies and advocacy groups in Europe, Japan, Korea, and elsewhere to accelerate the validation and adoption of other non-animal safety tests.

On a related front, much progress is being made to develop novel in vitro technologies composed of human tissues (vs. merely human cells) functioning in many of the same ways they do in vivo. Using the same microfabrication techniques developed to manufacture computer chips, scientists have created miniature versions of human organs that recapitulate many of the same functions, such as "lung-on-a-chip," "gut-on-a-chip," and, most recently, the entire menstrual cycle on a chip.[18] Another approach has been to culture various human stem cells under specific conditions so that they will self-assemble and function as "organoids," with many of the same properties and behaviors as if these stem cells matured into an organ inside the human body.[19] In addition to their applications in safety or toxicity testing, many envision these human tissue technologies for advancing our understanding about human disease and medicine as (1) a complement to and eventual replacement of animal models and (2) coming to the rescue for medical disciplines where progress has been impeded by a dearth of good animal models.[20] An even newer alternative to using laboratory animals for evaluating the efficacy of human drugs and vaccines is the use of pets that have the same or a similar disease as the target human patient population. Since these pets' afflictions are "natural" rather than induced, it is argued that they may represent a more predictive model as well as offer an opportunity for the pets themselves (and their owners) to benefit from such testing. This is a very recent development and raises as many ethical issues as well as tantalizing promises.[21]

While all this is going on, it is important that pertinent disciplines in the social sciences and humanities be recruited to monitor how societal attitudes about animal research change and in response to what inputs. Thankfully, analyses of this kind are available[22] and may offer fresh insights beyond the intransigence of polemics surrounding animal research established long ago. In addition, other developments beyond the obvious ones mentioned above in this chapter may influence how and how many laboratory animals are used in the future. One of the influences will be new knowledge about what animals actually think. For example, there is much evidence (as interpreted through human eyes) for the existence of naturally occurring empathy (as defined in human terms) in animals besides monkeys and apes.[23] In addition to being interesting on their own merits, such findings will likely encourage further claims of emotional equivalence between humans and animals, leading to an increase in objections to animal research on ethical grounds. If animals are found to be even more intelligent, sentient, and "feeling" in an anthropomorphic context, how will the use of laboratory animals be viewed a 100 years from now? Will it be universally cursed as a social evil on the same level as human slavery (which had its defenders in its day)? Conversely, what if we learn that animal minds are quite different rather than similar to ours? What if historical notions of "Nature, red in tooth and claw"[24] are corroborated by advances in comparative neuroscience so that animal behavior remains firmly hardwired to preservation of the organism and its progeny, destroying the cuddly, Disneyesque version of animals thinking and acting like four-legged persons? If so, would the use of animals in biomedical research and safety testing be interpreted in hindsight a 100 years from now to be morally acceptable as the best means available at the time to protect ourselves and our loved ones? And, therefore, would future ethicists conclude that animal activists should be blamed for impeding vital medical advances to overcome devastating diseases? These are simplistic questions and conjectures devoid of nuance, to be sure. But they deserve consideration alongside the continuing decline of our reliance on laboratory animals as the primary means to pursue the remaining scientific mysteries of human biology and medicine.

Around the turn of the 19th–20th century, there were 40 buggy whip companies in Westfield, Massachusetts alone (hence its nickname, "Whip City"), producing around three million whips a year at their peak.[25] US telegraph companies employed a maximum of 85,000 telegraphers according to the 1920 US census and transmitted over 236 million telegrams in 1945.[26] Today, few buggy whip firms remain in business, reduced to making whips for horse-drawn carriage and surrey hobbyists and professionals. Commercial transmission of telegrams was discontinued in the United States in 2006. Given everything described above and more that are replacing laboratory animals, are my successors destined to become the 21st century versions of 20th century buggy whip manufacturers and telegraph operators?

Perhaps. Scientific endeavors to learn about life on this planet will continue, including how animals function, interact, and respond to growing anthropogenic influences. But animal-based research will be markedly different than today's approach. Animals will be studied in less invasive ways. Rather than being injected or made to undergo surgery, animals will swallow or have applied to their skin nanotech sensors to identify changes in metabolism and disease in real time and display these changes via external devices involving temporary and comfortable restraint. Informative biological samples will be obtained more frequently from saliva, excreta, and shed hair rather than solely by venipuncture or requiring serial sacrifice. In order to understand how animals behave and live without the artificiality and interference of standardized confinement, they will be evaluated more often in their native habitats or in more naturalistic environments than today's vivarium can offer—maybe this will give zoos, aquaria, and wildlife refuges an expanded mission to participate in applied as well as basic biological research.

A recent example of such an expanded view of "research animals" and of which I'm personally familiar is how genes that drive mammalian behavior were identified for the first time. Observing how two species of deer mice, *Peromyscus maniculatus* and *P. polionotus*, differed in how they made burrows and raised their young in the wild, replicating the same behaviors in the rodent facility managed by my department, and then performing cross-breeding and employing modern genetic tools allowed scientists to determine which genes and their variants influenced how these animals behaved.[27] Wild-caught *Peromyscus* destined for our vivarium needed to be evaluated for murine and zoonotic pathogens prior to their arrival and then maintained in enclosures that permitted natural behaviors while not compromising animal well-being, worker safety, or regulatory compliance, or resulting in unreasonable impositions on vivarium operations. Facilitating such research within an SPF rodent barrier colony while protecting the rest of the resident population required cooperation and a willingness to try unconventional husbandry options.

It's also reasonable to anticipate that selected animals in non-vivarium settings will be allowed to acquire infection or develop disease "naturally" (instead of it being induced), and these animals will be followed intensively while some of them are administered experimental drugs or vaccines. Alternatively, the clinical trials standard of "do no harm" may become the ethical norm so that all afflicted animals, including those in control groups, are provided veterinary care, some with a conventional treatment regimen and others with the experimental version. And if either treatment arm fails to restore health, these animals will be euthanized via new and gentler methods sooner rather than allow them to worsen to obvious pain or distress before an endpoint is reached. On a related note, animals living in their conventional settings will become useful sentinels for how future advances in biotechnology will impact the environment, the economy, as well as human and animal health.[28] Wild animals and livestock will be monitored in nature and on farms for any effects on their welfare, production, and reproduction during experimental release of new genetically engineered crops, soil microbes, industrial enzymes, former insect vectors of disease, etc. While not as recognizable as today's laboratory animals, they are still research animals in the most basic sense of the concept.

What do these trends imply for laboratory animal care, medicine, and management? To quote from a US expert panel on anticipating disruptive technologies, "The value of technology forecasting lies not in its ability to accurately predict the future but rather in its potential to minimize surprises. It does this by various means:

- defining and looking for key enablers and inhibitors of new disruptive technologies,
- assessing the impact of potential disruption,
- postulating potential alternative futures, and
- supporting decision making by increasing the lead time for awareness."[29]

Can we predict macro-changes in the care, use, and popularity of laboratory animals? Even if that's unlikely, continuing to insist that everyone engaged in this field adhere to antiquated engineering standards and confine their views only to today's vivarium will ensure a stormy demise rather than a smooth evolution. What's needed, and starting now, is more flexibility, creativity, and curiosity, along the themes presented in this book. We should make ourselves known to electrical engineers, nano- and material chemists, software developers, and others to familiarize ourselves with emerging technologies to monitor and support animals in captivity. We should attend classes and seminars on trends in business management to modernize how we direct the various components of our operations at a lower cost and higher quality. And we should reach out to colleagues in other fields of zoology, such as wildlife conservation and animal science, to learn what they are investigating so that we may apply any resultant new knowledge on behalf of our customers.[30]

To do anything less would be most unfortunate.

# Endnotes

## Chapter 1

1. Fontana, L., Partridge, L., Longo, V.D., 2010. Extending healthy life span—from yeast to humans. Science 328, 321–326.
2. Lawler, D.F., Evans, R.H., Larson, B.T., Spitznagel, E.L., Ellersieck, M.R., Kealy, R.D., 2005. Influence of lifetime food restriction on causes, time, and predictors of death in dogs. J. Am. Vet. Med. Assoc. 226, 225–231.
3. Mattison, J.A., Colman, R.J., Beasley, T.M., Allison, D.B., Kemnitz, J.W., Roth, G.S., Ingram, D.K., Weindruch, R., de Cabo, R., Anderson, R.M., 2017. Caloric restriction improves health and survival of rhesus monkeys. Nat. Comm. 8, 14063. https://doi.org/10.1038/ncomms14063.
4. Ravussin, E., Redman, L.M., Rochon, J., Das, S.K., Fontana, L., Kraus, W.E., Romashkan, S., Williamson, D.A., Meydani, S.N., Villareal, D.T., Smith, S.R., Stein, R.I., Scott, T.M., Stewart, T.M., Saltzman, E., Klein, S., Bhapkar, M., Martin, C.K., Gilhooly, C.H., Holloszy, J.O., Hadley, E.C., Roberts, S.B., 2015. A 2-year randomized controlled trial of human caloric restriction: feasibility and effects on predictors of health span and longevity. J. Gerontol. A Biol. Sci. Med. Sci. 70, 1097–1104.
5. Haseman, J.K., Ney, E., Nyska, A., Rao, G.N., 2003. Effect of diet and animal care/housing protocols on body weight, survival, tumor incidences, and nephropathy severity of F344 rats in chronic studies. Toxicol. Pathol. 31, 674–681.
6. Rao, G.N., Crockett, P.W., 2003. Effect of diet and housing on growth, body weight, survival and tumor incidences of B6C3F1 mice in chronic studies. Toxicol. Pathol. 31 (2), 243–250.
7. https://en.wikipedia.org/wiki/Ad_libitum.
8. Stipp, D., 2013. Is fasting good for you? Sci. Am. 308, 23–24.
9. Grens, K., June 18, 2015. Periodic fasting improves rodent health. Science. (http://www.the-scientist.com/?articles.view/articleNo/43344/title/Periodic-Fasting-Improves-Rodent-Health/).
10. Swoap, S.J., 2008. The pharmacology and molecular mechanisms underlying temperature regulation and torpor. Biochem. Pharmacol. 76, 817–824.
11. Pritchett-Corning, K.R., Keefe, R., Garner, J.P., Gaskill, B.N., 2013. Can seeds help mice with the daily grind? Lab. Anim. 47 (4), 312–315.
12. Bauer, S.A., Arndt, T.P., Leslie, K.E., Pearl, D.L., Turner, P.V., 2011. Obesity in rhesus and cynomolgus macaques: a comparative review of the condition and its implications for research. Comp. Med. 61 (6), 514–526.
13. USDA APHIS Animal Care. Animal Welfare Regulations, Part 3, Subpart D, section 3.81(b).

## Chapter 2

1. https://en.wikipedia.org/wiki/Internet_of_Things.
2. Carr, A., October 2014. The $3.2 billion man: can Google's newest star outsmart Apple? Fast Company, Issue 189. http://www.fastcodesign.com/3035239/innovation-by-design-2014/nest-hatches-a-connected-home-boom.
3. The Next Big Thing, August 2002. The Pitch, A Newsletter for Professional Investors, San Francisco. www.the-pitch.com.

4. Even static cages that are not actively ventilated require fairly narrow environmental parameters, such as room temperature, humidity, and ambient light, all of which consume energy.
5. I wanted to find an official figure, but the US Department of Commerce, the US Census Bureau, and the US Department of Agriculture don't keep such statistics. And the leading refrigerator manufacturers were similarly uninformative. So I estimated what a usual trip to the supermarket costs, less the goods that don't require refrigeration or freezing, and figured this was close enough for the purpose of this chapter. The retail cost of the groceries nicely comprises all of their production costs, including any publicly funded subsidies and protective tariffs for domestic producers, as well as transportation to market and the retail outlet's operating costs for making these items available for consumer purchase with whatever profits taken by each segment of the supply chain.
6. US Patent 2957320 A, "Refrigerating apparatus with magnetic door seal actuated switch," Everett C Armentrout, Inventor (Publication date October 25, 1960; Filing date September 28, 1959). This patent mentions earlier door-ajar alarms of a spring-pressed plunger type in refrigerators that are improved by replacing the springs with magnets. It doesn't matter if this is not the first ever mention of door-ajar alarms on refrigerator doors. The mere fact that this "advance" was invented 56 years ago should be sufficiently embarrassing as a contrast to how rodent caging has not similarly progressed.
7. No figures have been published or even mentioned in passing of which I am aware. So I started with figures for a prominent academic laboratory I know that's dependent on externally sourced grants for life sciences research focused almost solely on mice for all of its investigations and findings. All direct costs in this laboratory, including wages and salaries plus fringe benefits, procured animals, reagents and other consumable supplies, non-capital equipment, miscellaneous allowable expenses such as travel to scientific conferences, etc. were burdened with an internal overhead rate of 84% for infrastructure, the result of which totaled $14.2 million over the past 5 years. This laboratory was responsible for slightly over 1.8 million mouse cage days over the same time period, so each cage of mice represented $7.87 in tangible research investments on any given day, including fully burdened (including indirect/overhead) costs of $1.40 per day for routine husbandry and veterinary support. These figures also indicate that the daily sunk costs in research alone were almost five times the cost of merely keeping the mice comfortable and healthy ($6.47 versus $1.40). Without question, more data and analyses than a single sampling are needed and I encourage others to investigate this topic further.
8. "Digital Ventilated Cages," by Tecniplast SpA, Buguggiate, Italy. http://www.tecniplast.it/en/dvc.html.
9. Augenbraun, E., June 18, 2014. First scented messages sent across the Atlantic, CBS News. http://www.cbsnews.com/news/first-scented-message-sent-across-the-atlantic/.
10. Holy, T.E., Guo, Z., 2005. Ultrasonic songs of male mice. PLoS Biol. 3 (12), e386.
11. "Rodent ultrasonic sounds as an indicator of nonevoked, postoperative pain" by J. Borzan, presented at the ACLAM Forum, Williamsburg, VA, April 17, 2013.
12. Arts, J.W.M., Kramer, K., Arndt, S.S., Ohl, F., 2012. The impact of transportation on physiological and behavioral parameters in Wistar rats: implications for acclimatization periods. ILAR J 53 (1), E82–E98; Stemkens-Sevens, S., van Berkel, K., de Greeuw, I., Snoeijer, B., Kramer, K., 2009. The use of radiotelemetry to assess the time needed to acclimatize guinea pigs following several hours of ground transport. Lab. Anim. 43, 78–84.
13. Rodrigue, J.-P., Notteboom, T., 2013. The cold chain and its logistics. In: Rodrigue J.-P. (Eds.), The Geography of Transport Systems, Routledge, New York.
14. Excerpted from "The Value of Fresh Air: New Approaches to In-shipment Monitoring," a presentation I delivered at a session on laboratory animal transportation at the American Veterinary Medical Association 2015 Annual Convention, Boston, MA, July 12. The mentioned companies

that make simple environmental monitoring devices for use in animal shipping crates include HOBO Date Loggers by Onset Computer Corporation and LogTag by LogTag Recorders Limited.

15. They already are, in most programs, as are veterinarians. But with remote cage sensors able to activate alarms at any time, it's likely that middle managers will receive more calls than ever before and some sharing or relief is indicated, which could require hiring more supervisors.
16. Anon., November 21, 2015 Smart Products, Smart Makers. The Economist. http://www.economist.com/news/business-and-finance/21678748-old-form-capitalism-based-built-obsolescence-giving-way-new-one-which.
17. Siegel, E., 2016. Predictive Analytics: The Power to Predict Who Will Click, Buy, Lie, or Die. John Wiley & Sons, Hoboken, NJ.
18. Davenport, T.H., January 14, 2015. The rise of automated analytics. The CIO Report. Wall Str. J. (http://blogs.wsj.com/cio/2015/01/14/the-rise-of-automated-analytics/).
19. When bar coding replaced manual counting and recording, we were able to reduce the time interval between when the cage was counted to when the invoice was actually sent to investigators for charging per diems to their grants from 2 months to 2 days! (Jarrell, D.M., Moore, M.J., Beck, B.N., Young, S.H., Sun, Y., Taylor, S., Camera, L., La Mattina-Smith, N., Niemi, S.M., 2004. The use of bar-coding and PDA scanners in managing an animal facility. Poster #092, 55th Annual Session, American Association for Laboratory Animal Science. Oct 17–20, 2004, Tampa, FL).
20. Beck, B.N., Jarrell, D.M., Young, S.H., Niemi, S.M., 2004. Initial application of RFID technology in the vivarium. Abstract #51, 55th Annual Session, American Association for Laboratory Animal Science. Oct 17–20, 2004, Tampa, FL.
21. Instead of squeezing a commercial smart cage firm by insisting on a free trial, at least offer to pay for transport and installation costs. Vendors need to stay profitable to stay in business, and we all benefit from financially healthy suppliers.

## Chapter 3

1. Kokolus, K.M., Capitano, M.L., Lee, C.-T., Eng, J.W.-L., Waight, J.D., Hylander, B.L., Sexton, S., Hong, C.-C., Gordon, C.J., Abrams, S.I., Repasky, E.A., 2013. Baseline tumor growth and immune control in laboratory mice are significantly influenced by subthermoneutral housing temperature. Proc. Natl. Acad. Sci. 110 (50), 20176–20181.
2. Grant, B., 2015. Brrrr-ying the results. Sci. 29 (7), 16–17; Dvorsky, G., April 19,2016. Lab Mice are Freezing Their Asses Off – and That's Screwing Up Science. Gizmodo Media Group. http://gizmodo.com/lab-mice-are-freezing-their-asses-off-and-that-s-screwi-1771796664.
3. Willmer, P., Stone, G., Johnston, I. 2005. Environmental Physiology of Animals, second ed. Oxford, Blackwell, pp. 40–41.
4. Gordon, C.J., 2012. Thermal physiology of laboratory mice: defining thermoneutrality. J. Thermal Biol. 37, 654–685.
5. Swoap, S.J., Overton, J.M., Garber, G., 2004. Effect of ambient temperature on cardiovascular parameters in rats and mice: a comparative approach. Am. J. Physiol. Regul. Integr. Comp. Physiol. 287, 391–396.
6. Chen, K.K., Anderson, R.C., Steldt, F.A., Mills, C.A., 1943. Environmental temperature and drug interaction in mice. J, Pharmacol, Exp. Ther. 79 (2), 127–132.
7. Gaskill, B.N., Rohr, S.A., Pajor, E.A., Lucas, J.R., Garner, J.P., 2011. Working with what you've got: changes in thermal preference and behavior in mice with or without nesting material. J. Therm. Biol. 36, 193–199; Gaskill, B.N., Gordon, C.J., Pajor, E.A., Lucas, J.R., Davis, J.K., Garner, J.P., 2013 Impact of nesting material on mouse body temperature and physiology. Physiol. Behav. 110–111, 87–95.
8. Gaskill, B.N., Karas, A.Z., Garner, J.P., Pritchett-Corning, K.R., 2013. Nest building as an indicator of health and welfare in laboratory mice. J. Vis. Exp. 82, 51013.

9. Chance, M.R.A., 1943. Factors influencing the toxicity of sympathomimetic amines to solitary mice. J. Pharmacol. Exp. Ther. 89 (3), 289–296.

10. Helppi, J., Schreier, D., Naumann, R., Zierau, O., 2016. Mouse reproductive fitness is maintained up to an ambient temperature of 28°C when housed in individually-ventilated cages. Lab. Anim. 50 (4), 254–263.

11. David, J.M., Knowles, S., Lamkin, D.M., Stout, D.B., 2013. Individually ventilated cages impose cold stress on laboratory mice: a source of systemic experimental variability. J. Am. Assoc. Lab. Anim. Sci. 52, 738–744.

12. Crowley, M.A., Sedlacek, R.M., Niemi, S.M., 2016. A novel device for locally heating ventilated rodent cages. 66th National Meeting of the American Association for Laboratory Animal Science, Phoenix, AZ, November 1–5, 2016, Poster 182.

13. http://www.altdesign.com/solacezone/.

14. Speakman, J.R., Keijer, J., 2013. Not so hot: optimal housing temperatures for mice to mimic the thermal environment of humans. Mol. Metab. 2, 5–9.

## Chapter 4

1. Walker, C., Streisinger, G., 1983. Induction of mutations by gamma-rays in pregonial germ cells of zebrafish embryos. Genetics 103 (1), 125–236.

2. Source: PubMed (http://www.ncbi.nlm.nih.gov/pubmed/), using "zebrafish" as the keyword for All Fields, January 15, 2016.

3. I had proposed calling it the Boston Zebrafish Organization, instead, with the intentionally eye-catching acronym "BOZO." Thankfully, I was politely ignored. Within just a couple of years, "New England" was dropped from the name because so many laboratory zebrafish care experts from around the world had joined. So it became simply the Zebrafish Husbandry Association (www.zhaonline.org). NEZHA's founding is also noted in Baur, B.M., 2010. Zebrafish: the challenge of standardization in a maturing research model. Lab Anim News. http://www.aln-mag.com/articles/2010/05/zebrafish-challenge-standardization-maturing-research-model.

4. Papers on the reliability of rodent microisolation cages to protect their inhabitants from nearby infection span several decades, from Lipman, N.S., Newcomer, C.E., Fox, J.G., 1987. Rederivation of MHV and MEV antibody positive mice by cross-fostering and use of the microisolator caging system. Lab. Anim. Sci. 37 (2), 195–199, to Baker, S.W., Prestia, K.A., Karolewski, B. 2014. Using reduced personal protective equipment in an endemically infected mouse colony. J. Am. Assoc. Lab. Anim. Sci. 53 (3), 273–277.

5. Kent, M.L., Feist, S.W., Harper, C., Hoogstraten-Miller, S., Mac Lawe, J.M., Sánchez-Morgado, J.M., Tanguay, R.L., Sanders, G.E., Spitsbergen, J.M., Whipps, C.M., 2009. Recommendations for control of pathogens and infectious diseases in fish research facilities. Comp. Biochem. Physiol. Part C Toxicol. Pharmacol. 149, 240–248.

6. Chris Lawrence interview, Boston, November 20, 2015.

7. Christopher, M., Whipps and Michael L. Kent., 2006. Polymerase chain reaction detection of *Pseudoloma neurophilia*, a common microsporidian of zebrafish (*Danio rerio*) reared in research laboratories. J. Amer. Assoc. Lab. Anim. Sci. 45 (1), 36–39.

8. Spagnolia, S., Xue, L., Kent, M.L., 2105. The common neural parasite *Pseudoloma neurophilia* is associated with altered startle response habituation in adult zebrafish (*Danio rerio*): implications for the zebrafish as a model organism. Behav. Brain. Res. 291, 351–360; Spagnoli, S.T., Xue, L., Murray, K.N., Chow, F., Kent, M.L., 2015. *Pseudoloma neurophilia*: A retrospective and descriptive study of nervous system and muscle infections, with new implications for pathogenesis and behavioral phenotypes. Zebrafish 12 (2), 189–201.

9. Kent, M.L., Harper, C., Wolf, J.C., 2102. Documented and potential research impacts of sub-clinical diseases in zebrafish. ILAR J. 53 (2), 126–134.

10. Murray, K.N., Dreska, M., Nasiadka, A., Rinne, M., Matthews, J.L., Carmichael, C., Bauer, J., Varga, Z.M., Westerfield, M. 2011. Transmission, diagnosis, and recommendations for control of *Pseudoloma neurophilia* infections in laboratory zebrafish (*Danio rerio*) facilities. Comp. Med. 61 (4), 322–329.

11. Kent, M.L., Buchner, C., Watral, V.G., Sanders, J.L., Ladu, J., Peterson, T.S., Tanguay, R.L. 2011. Development and maintenance of a specific pathogen-free (SPF) zebrafish research facility for *Pseudoloma neurophilia*. Dis. Aquat. Organ. 95 (1), 73–79.

12. Lawrence, C., Ennis, D.G., Harper, C., Kent, M.L., Murray, K., Sanders, G.E., 2012. The challenges of implementing pathogen control strategies for fishes used in biomedical research. Comp. Biochem. Physiol. C Toxicol. Pharmacol. 155 (1), 160–166.

13. Kelly, S.O., 2015. Disease detector. Sci. Am. 313 (5), 49–51.

14. Shannon, M.A., Bohn, P.W., Elimelech, M., Georgiadis, J.G., Mariñas, B.J., Mayes, A.M., 2008. Science and technology for water purification in the coming decades. Nature 452, 301–310; Zhang, M., Xie, X., Tang, M., Criddle, C.S., Cui, Y., Wang, S.X., 2013. Magnetically ultraresponsive nanoscavengers for next-generation water purification systems. Nat. Comm. 4, 1866. https://doi.org/10.1038/ncomms2892.

15. Linh, N., Pham, L.N., Kanther, M., Semova, I., Rawls, J.F., 2008. Methods for generating and colonizing gnotobiotic zebrafish. Nat. Protoc. 3 (12), 1862–1875.

16. Rawls, J.F., Mahowald, M.M., Ley, R.E., Gordon, J.I., 2006. Reciprocal gut microbiota transplants from zebrafish and mice to germ-free recipients reveal host habitat selection. Cell 127, 423–433; McFall-Ngai, M.M., 2006. Love the one you're with: vertebrate guts shape their microbiota. Cell 127, 247–249.

## Chapter 5

1. Clark, J.D., Gebhart, G.F., Gonder, J.C., Keeling, M.E., Kohn, D.F., 1997. The 1996 Guide for the Care and Use of Laboratory Animals. ILAR J. 38 (1), 41–48.

2. https://grants.nih.gov/grants/olaw/positionstatement_guide.htm.

3. https://grants.nih.gov/grants/olaw/references/phspol.htm#2.

4. Some have suggested that even longer intervals between editions of the Guide are even better, and there is also a valid rationale that the eighth edition should be the last. That's because any capital investment necessitated by the last revision be given sufficient time for extended use and depreciation before the next edition creates a need to replace that equipment with something compliant with any new engineering standards. Thus, there is an argument to be made that the eighth edition of the Guide should be the final one, or at least for a long time.

5. NIH Research Funding Trends. Federation of American Societies for Experimental Biology. http://faseb.org/Science-Policy-and-Advocacy/Federal-Funding-Data/NIH-Research-Funding-Trends.aspx. These numbers exclude a one-time boost provided by the American Recovery and Reinvestment Act of 2009 that was restricted to research infrastructure.

6. 2015 Profile—Biopharmaceutical Industry. Washington: Pharmaceutical Research and Manufacturers of America, p. 36. http://phrma-docs.phrma.org/sites/default/files/pdf/2015_phrma_ profile.pdf.

7. Baker, S.W., Prestia, K.A., Karolewski, B., 2014. Using reduced personal protective equipment in an endemically infected mouse colony. J. Am. Assoc. Lab. Anim. Sci. 53 (3), 273–277.

8. Lipman, N.S., Newcomer, C.E., Fox, J.G., 1987. Rederivation of MHV and MEV antibody positive mice by cross-fostering and use of the microisolator caging system. Lab. Anim. Sci. 37 (2), 195–199; Lipman, N.S., Corning, B.F., Saifuddin, M., 1993. Evaluation of isolator caging systems for protection of mice against challenge with mouse hepatitis virus. Lab. Anim. 27 (2), 134–140; Macy, J.D., Cameron, G.A., Ellis, S.L., Hill, E.A., Compton, S.R., 2002. Assessment of static isolator cages with automatic watering when used with conventional husbandry techniques as a factor in the transmission of mouse hepatitis virus. Contemp. Top. Lab. Anim. Sci. 41 (4), 30–35.

9. Gordon, S., Fisher, S.W., Raymond, R.H., 2001. Elimination of mouse allergens in the working environment: assessment of individually ventilated cage systems and ventilated cabinets in the containment of mouse allergens. J. Allergy Clin. Immunol. 108 (2), 288–294; Thulin, H., Björkdahl, M., Karlsson, A.S., Renström, A., 2002. Reduction of exposure to laboratory animal allergens in a research laboratory. Ann. Occup. Hyg. 46 (1), 61–68; Feistenauer, S., Sander, I., Schmidt, J., Zahradnik, E., Raulf, M., Brielmeier, M., 2014. Influence of 5 different caging types and the use of cage-changing stations on mouse allergen exposure. J. Am. Assoc. Lab. Anim. Sci. 53 (4), 356–363.

10. Many insist that because 180°F/82.2°C is also stipulated in the AWA Regulations for sanitizing primary enclosures (e.g., 9 CFR, Chapter 1, Subchapter A, Part 3, Subpart A, §3.11(b)(3)(ii) for dogs and cats), it must still be followed for regulated species. However, the very next clause following this engineering (water temperature) standard, wherever it's printed, offers a performance standard as an alternative. For primary enclosures housing dogs and cats, it's stated as "Washing all soiled surfaces with appropriate detergent solutions and disinfectants, or by using a combination detergent/disinfectant product that accomplishes the same purpose, with a thorough cleaning of the surfaces to remove organic material, so as to remove all organic material and mineral buildup, and to provide sanitization followed by a clean water rinse." (Subpart A, §3.11(b)(3)(iii)). So, contrary to conventional thought, there are options for cleaning cages of even AWA-covered species after all.

11. Sedlacek, R.S., Huang, P., Niemi, S.M., 2005. Significant energy conservation using cold water for washing rodent cages in a gnotobiotic facility. 56th Annual Meeting, American Association for Laboratory Animal Science, Saint Louis, MO, Abstract #PS40; Niemi, S.M., Sedlacek, R.S., Alves, H., Jarrell, D.M., 2009; Clean, Green, and Lean: Innovative Approaches to Rodent Barrier Management. Special seminar session, 60th Annual Meeting, American Association for Laboratory Animal Science, Denver, CO.

12. In my current program, we found that in rodent rooms doubly supplied with high-efficiency particulate arrest–filtered air, no residue or even dust was detectable on the ceilings after 9 years of occupancy! It's a good thing I didn't have blinders on and insist that the ceilings be washed according to a schedule. If and when they ever look dusty or come up with higher levels of residue, then and only then will they be cleaned.

13. Rosenbaum, M.D., VandeWoude, S., Volckens, J., Johnson, T.E., 2010. Disparities in ammonia, temperature, humidity, and airborne particulate matter between the micro- and macroenvironments of mice in individually ventilated caging. J. Am. Assoc. Lab. Anim. Sci. 49 (2), 177–183.

14. IVC cages are often changed at least every 2 weeks rather than one because the increased air exchange inside these cages delays buildup of ammonia from excreta. This allows them to go longer than static cages before they need changing.

15. Brown, P., 2009. Extending the cage interval period for CD-1 mice: are there welfare implications? Anim. Technol. Welfare 8 (2), 39–48; Reeb-Whitaker, C.K., Paigen, B., Beamer, W.G., Bronson, R.T., Churchill, G.A., Schweitzer, I.B., Myers, D.D., 2006. The impact of reduced frequency of cage changes on the health of mice housed in ventilated cages. Lab. Anim. 40, 353–370.

16. Brandolini, J., Gentile, S., Pina, F., Jarrell, D.M., 2009. Standardizing the definition of a "dirty" soiled rodent cage. 60th Annual Meeting, American Association for Laboratory Animal Science, Denver CO, Abstract #30; Wiler, R., 2012. Reducing waste via an innovative cage change process in a GEMM production facility. ALN Mag. https://www.alnmag.com/article/2012/10/reducing-waste-innovative-cage-change-process-gemm-production-facility; Washington, I.M., Payton, M.E., 2016. Ammonia levels and urine-spot characteristics as cage-change indicators for high-density individually ventilated mouse cages. J. Am. Assoc. Lab. Anim. Sci. 55 (3), 1–8.

17. Rasmussen, S., Miller, M.M., Filipski, S.B., Tolwani, R.J., 2011. Cage change influences serum corticosterone and anxiety-like behaviors in the mouse. J. Am. Assoc. Lab. Anim. Sci. 50 (4), 479–483 (NB: cited references in this paper include other published studies that show lower cage changing frequencies or better informed timing of cage changes has fewer or no adverse effects).

18. Guidance on Significant Changes to Animal Activities. NIH Office of Laboratory Animal Welfare Notice NOT-OD-14-126. https://grants.nih.gov/grants/guide/notice-files/NOT-OD-14-126. html.

19. One hypothetical compliance shortcut that intrigues me and appears entirely legitimate at first glance is to eliminate the de novo triennial rewrite and review of USDA protocols that are not PHS funded. Since USDA requires an annual IACUC review of protocols using species covered by the AWA, why couldn't that annual review suffice for OLAW?

20. Jensen, E.S., Allen, K.P., Henderson, K.S., Szabo, A., Thulin, J.D., 2013. PCR testing of a ventilated caging system to detect murine fur mites. J. Am. Assoc. Lab. Anim. Sci. 52 (1,28–33; Henderson, K.S., Perkins, C.L., Havens, R.B., Kelly, M.J., Francis, B.C., Dole, V.S., Shek, W.R., 2013. Efficacy of direct detection of pathogens in naturally infected mice by using a high-density PCR array. J. Am. Assoc. Lab. Anim. Sci. 52 (6), 763–772.

21. Lauer, M., 2017. FY2016 by the numbers. NIH Extramural Nexus. https://nexus.od.nih.gov/all/2017/02/03/fy2016-by-the-numbers/.

22. Readers are welcome to calculate their own percentages, using internal financial data from any US academic research institution(s) of interest. My expectation is that for large and established institutions engaged in a typical mix of research for clinical (patient) and/or non-biomedical endeavors, the portion of the entire research budget spent on direct (operating) expenses for laboratory animal care will fall in the 1%–2% range so my calculations and conclusions are expected to remain accurate enough to be applicable across the country. However, further exploration of these estimates is both merited and welcomed.

23. Rocky, S., March 25, 2015. More data on age and the workforce. Extramural Nexus, NIH Office of Extramural Research. https://nexus.od.nih.gov/all/2015/03/25/age-of-investigator/.

24. Lauer, M., November 4, 2016. R01 and R21 applications & awards: trends and relationships across NIH. Extramural Nexus, NIH Office of Extramural Research. https://nexus.od.nih.gov/all/2016/11/04/nih-r01-r21/.

25. Kaiser, J., December 15, 2016. NIH discusses curbing lab size to fund more midcareer scientists. Science Insider. http://www.sciencemag.org/news/2016/12/nih-discusses-curbing-lab-size-fund-more-midcareer-scientists.

26. Table #103, NIH Research Grants, Awards and Total Funding by Grant Mechanism and Activity Code, Fiscal Years 2007–16, Office of Extramural Research, National Institutes of Health, extracted from https://report.nih.gov/catalog.aspx.

27. DiMaggio, P.J., Powell, W.W., 1983. The iron cage revisited: institutional isomorphism and collective rationality in organizational fields. Am. Sociol. Rev. 48 (2), 147–160.

28. Such extramural entities include not only USDA, OLAW, and AAALAC but also the various organizations in the United States through which individuals enter the field and build their professional careers, such as the AALAS, ACLAM, ASLAP, PRIM&R, and SCAW.

29. Niemi, S.M., April 21, 2015. Sharing of acceptable performance standards. ILAR Roundtable Workshop – Design, Implementation, Monitoring and Sharing of Performance Standards, Washington DC. Available at: http://nas-sites.org/ilar-roundtable/roundtable-activities/performance-standards/.

30. No regulators or anyone officially representing an accreditation agency would be allowed to serve on filter panels. Even though they may be qualified and knowledgeable, they'd either feel compelled to weigh in on the compatibility of posted content with their respective organizations, or they'd risk compromising their official roles by appearing to "approve" a submission for posting that may conflict with their regulatory or accreditation standards. For the same reasons, no postings would be permitted from anyone who comments on the regulatory or accreditative acceptability of a submission or subsequent annotations. Even if a submission doesn't comply with US laws and regulations, it may be allowed under another jurisdiction and, therefore, should still be posted for all to digest. Readers could then decide if there's a compliance problem for their respective institutions.

31. Anon., 2013. Vertebrate of the Year – The rat that ages beautifully. Science 342, 1435.

32. Buffenstein, R., Park, T., Hanes, M., Artwohl, J.E., 2012. Naked mole rat, in The Laboratory Rabbit, Guinea Pig, Hamster, and Other Rodents. Suckow, M.A., Stevens, K.A., Wilson, R.P. (Eds), New York, Elsevier Inc., pp 1055–1074; Yu, C., Wang, S., Yang, G., Zhao, S., Lin, L., Yang, W., Tang, Q., Sun, W., Cui, S., 2017. Breeding and rearing naked mole-rats (*Heterocephalus glaber*) under laboratory conditions. J. Am. Assoc. Lab. Anim. Sci. 56 (1), 98–101.

33. Storrs, E.E., Greer, W.E., 1973. Maintenance and husbandry of armadillo colonies. Lab. Anim. Sci. 23 (6), 823–829.

34. Reardon, S., 2016. Dirty room-mates make lab mice more useful. Nature 532, 294–295.

35. Henig, R.M., 2016. From research to reward: the hospital checklist: how social science insights improve health care outcomes. National Academies Press. Available at: http://www.nap.edu/23510; Bender, E., 2016. Crowdsourced solutions. Nature 533, S62–S64; Nanos, J., October 28, 2016. Designers bring private-sector ideas to public policy. Boston Globe; IEEE (originally, Institute of Electrical and Electronics Engineers) Collabratec ("an integrated online community where technology professionals can network, collaborate, and create—all in one central hub"). http://ieee-collabratec.ieee.org/.

36. Tapscott, D., Williams, A.D., 2008. Wikinomics. How Mass Collaboration Changes Everything, Penguin, London; Leading Open Innovation. Huff, A.S., Moslein, K.M., Reichwald, R. (Eds), August 2013. MIT Press Scholarship Online. https://doi.org/10.7551/mitpress/9780262018494.001.0001.

## Chapter 6

1. Bennett, M.K., 1941. Wheat in national diets. Wheat Studies of the Food Research Institute 18 (2), 37–76; Périssé, J., Sizaret, F., Francois, P., 1969. The effect of income on the structure of the diet. Nutr. Newsl. 7 (3), 1–9; Timmer, C.P., Falcon, W.P., Pearson, S.R., 1983. Food Policy Analysis. The Johns Hopkins University Press, Baltimore.

2. Joint WHO/FAO Expert Consultation on Diet, Nutrition and the Prevention of Chronic Diseases, 2003. Diet, nutrition and the prevention of chronic diseases: report of a joint WHO/FAO expert consultation, Geneva, January 28–February 1, 2002. WHO Technical Report Series No. 916; Delgado, C.L., 2003. Rising consumption of meat and milk in developing countries has created a new food revolution. J. Nutr. 133, 3907S–3910S; Kearney, J., 2010. Food consumption trends and drivers. Philos. Trans. R Soc. B 365, 2793–2807; Gandhi, V.P., Zhou, Z.-Y., 2010. Rising demand for livestock products in India: nature, patterns and implications. Australasian Agribus. Rev. 18, Paper 7, ISSN: 1442-6951.

3. Animal Health India. http://www.animalhealthindia.com/pet-animal-population.

4. Pet Care Forecast Revisit 2012: How Resilient Is The Global Market? Euromonitor International, 2012.

5. Katsnelson, A., Mcdonald, A., 2016. Big science spenders. Nature 537, S2–S3.

6. GERD (gross Domestic Expenditure on R&D), million constant USD (United States dollars) PPPs (Purchasing Power Parities), https://www.innovationpolicyplatform.org/content/statistics-ipp.

7. For example, June 22, 2016. Science in China. Nature. http://www.nature.com/news/science-in-china-1.20094; May 13, 2015. Science in India. Nature. http://www.nature.com/news/india-1.17456.

8. Cyranoski, D., 2016. Monkey Kingdom. Nature 532, 300–302.

9. Many thanks to Chris Newcomer and AAALAC International for generously providing the geographical data from which these calculations were made.

10. Accredited Organizations. AAALAC International. https://www.aaalac.org/accreditedorgsdirectorysearch/aaalaclistall.cfm.

11. Institute for Laboratory Animal Research, 2011. Animal Research in a Global Environment: Meeting the Challenges: Proceedings of the November 2008 International Workshop. National Academies Press, Washington.

12. Bayne, K., Bayvel, D., MacArthur Clark, J., Demers, G., Joubert, C., Kurosawa, T.M., Rivera, E., Souilem, O., Turner, P.V., 2011. Harmonizing veterinary training and qualifications in laboratory animal medicine: a global perspective. ILAR J. 52 (3), 393–403.

13. Bayne, K., Ramachandra, G.S., Rivera, E.A., Wang, J., 2015. The evolution of animal welfare and the 3Rs in Brazil, China, and India. J. Am. Assoc. Lab. Anim. Sci. 54 (2), 181–191.

14. McLaughlin, K., March 21, 2016. China finally setting guidelines for treating lab animals. Science, posted in Asia/Pacific Policy. https://doi.org/10.1126/science.aaf4021.

15. Favoretto, S., Araújo, D., Oliveira, D., Duarte, N., Mesquita, F., Zanotto, P., Durigon, E., 2016. First detection of Zika virus in neotropical primates in Brazil: a possible new reservoir. bioRxiv preprint first posted online Apr. 20, 2016. https://doi.org/10.1101/049395.

16. Mysorekar, I.U., Diamond, M.S., 2016. Modeling Zika Virus Infection in Pregnancy. N. Engl. J. Med. 375 (5), 481–484.

17. Worley, K.C., et al., 2014. The common marmoset genome provides insight into primate biology and evolution. Nat. Genet. 46 (8), 850–857; Willyard, C., 2016. New models: gene-editing boom means changing landscape for primate work. Nat. Med. 22 (11), 1200–1202.

18. Wooldridge, A., April 15, 2010. The world turned upside down. The Economist (this introduction is followed by a series of articles in the same issue that analyzed innovation in LMICs).

19. Doland, A., 2014. Purina navigates China's complex market where owners often cook for their pampered pets. Marketer insight on how the one-child policy affects pet ownership. Advertising Age. http://adage.com/article/news/purina-navigates-china-s-market-owners-cook-pets/296085/.

20. Pet Humanisation: The Trend and Its Strategic Impact on Global Pet Care Markets. Euromonitor International, 2014; Entis, L., 2016. Pets are basically people. Fortune. http://fortune.com/2016/09/07/pets-are-basically-people/.

## Chapter 7

1. These appointments are similar, if not identical, to clinical residencies, but the latter title is not used in my specialty. That's because our training is supposed to be more comparable to post-doctoral research appointments for scientists who have just earned their PhDs, rather than veterinarians who focus on treating sick pets or livestock. But clinical residents are expected to perform original research that's to be published, too. So the rationale for the distinction in titles has never been clear to me.

2. https://www.avma.org/ProfessionalDevelopment/Education/Specialties/.

3. https://www.avma.org/Events/Symposiums/TheConversation.

4. http://www.sheltervet.org/.

5. In addition, laboratory animal care programs may have consumables that have expired, according to the manufacturer's label but are otherwise just as safe and efficacious. This is not to suggest that pet shelters use outdated materials and subject their animals to unnecessary risks, or encourage them to do so. But in some cases, laboratory animal regulatory obedience to expiration dates can be downright silly—think of surgical gauze or plastic syringes that exceed their suppliers' shelf life by 1–2 months. We have to throw these items out while someone else in a less-regulated environment could still use them.

## Chapter 8

1. For the sake of brevity and easier reading, only human medicine and human patients will be featured in this chapter, even though there is just as much reason to start with current veterinary care practices in other fields to see what could be reasonably transferred to laboratory animal medicine, independent of research aims and procedures for which these animals are used.

2. Kutter, A.P.N., Mauch, J.Y., Riond, B., Martin-Jurado, O., Spielmann, N., Weiss, M., Bettschart-Wolfensberger, R., 2012. Evaluation of two devices for point-of-care testing of haemoglobin in neonatal pigs. Lab. Anim. 46 (1), 65–70; Clemmons, E.A., Stovall, M.I., Owens, D.C., Scott, J.A., Jones-Wilkes, A.C., Kempf, D.J., Ethun, K.F., 2016. Accuracy of human and veterinary point-of-care glucometers for use in rhesus macaques (*Macaca mulatta*), sooty mangabeys (*Cercocebus atys*), and chimpanzees (*Pan troglodytes*). J. Am. Assoc. Lab. Anim. Sci. 55 (3), 346–353.

3. Vashist, S.K., Luppa, P.B., Yeo, L.Y., Ozcan, A., Luong, J.H.T., 2015. Emerging technologies for next-generation point-of-care testing. Trends Biotechnol. 33 (11), 692–705; Contreras-Naranjo, J.C., Wei, Q., Ozcan, A., 2016. Mobile phone-based microscopy, sensing, and diagnostics. IEEE J. Sel. Topics Quantum Electron. 22 (3). https://doi.org/10.1109/JSTQE.2015.2478657; Pardee, K., Green, A.A., Takahashi, M.K., Braff, D., Lambert, G., Lee, J.W., Ferrante, T., Ma, D., Donghia, N., Fan, M., Daringer, N.M., Bosch, I., Dudley, D.M., O'Connor, D.H., Gehrke, L., Collins, J.J., 2016. Rapid, low-cost detection of Zika virus using programmable biomolecular components. Cell 165, 1255–1266.

4. Jones, C.N., Hoang, A.N., Martel, J.M., Dimisko, L., Mikkola, A., Inoue, Y., Kuriyama, N., Yamada, M., Hamza, B., Kaneki, M., Warren, H.S., Brown, D.E., Irimia, D., 2016. Microfluidic assay for precise measurements of mouse, rat, and human neutrophil chemotaxis in whole-blood droplets. J. Leukoc. Biol. 100 (1), 241–247. Irimia, D., Ellett, F., 2016. Big insights from small volumes: deciphering complex leukocyte behaviors using microfluidics. J. Leukoc. Biol. 100 (2), 291–304.

5. Relling, M.V., Evans, W.E., 2015. Pharmacogenomics in the clinic. Nature 526, 343–350.

6. Mealey, K.L., Meurs, K.M., 2008. Breed distribution of the ABCB1-1Δ (multidrug sensitivity) polymorphism among dogs undergoing ABCB1 genotyping. JAVMA 233 (6), 921–924.

7. Tegeder, I., Costigan, M., Griffin, R.S., Abele, A., Belfer, I., Schmidt, H., Ehnert, C., Nejim, J., Marian, C., Scholz, J., Wu, T., Allchorne, A., Diatchenko, L., Binshtok, A.M., Goldman, D., Adolph, J., Sama, S., Atlas, S.J., Carlezon, W.A., Parsegian, A., Lötsch, J., Fillingim, R.B., Maixner, W., Geisslinger, G., Max, M.B., Woolf, C.J., 2006. GTP cyclohydrolase and tetrahydrobiopterin regulate pain sensitivity and persistence. Nat. Med. 12 (11), 1269–1277.

8. Trescot, A.M., Faynboym, S., 2014. A review of the role of genetic testing in pain medicine. Pain Physician 17 (5), 425–445; Trescot, A.M., 2014. Genetics and implications in perioperative analgesia. Best Pract. Res. Clin. Anaesthesiol. 28 (2), 153–166; Seripa, D., Latina, P., Fontana, A., Gravina, C., Lattanzi, M., Savino, M., Gallo, A.P., Melchionda, G., Santini, S.A., Margaglione, M., Copetti, M., di Mauro, L., Panza, F., Greco, A., Pilotto, A., 2015. Role of CYP2D6 polymorphisms in the outcome of postoperative pain treatment. Pain Med. 16 (10), 2012–2023.

9. MacLeod, A.K., McLaughlin, L.A., Henderson, C.J, Wolf, C.R., 2017. Application of mice humanised for cytochrome P450 CYP2D6 to the study of tamoxifen metabolism and drug-drug interaction with antidepressants. Drug Metab. Dispos. 45 (1), 17–22.

10. Sarwal, M., Chua, M.-S., Kambham, N., Hsieh, S.-C., Satterwhite, T., Masek, M., Salvatierra, O., 2003. Molecular heterogeneity in acute renal allograft rejection identified by DNA microarray profiling. N. Engl. J. Med. 349, 125–138.

11. U.S. National Library of Medicine. Number of authors per MEDLINE®/PubMed® citation. https://www.nlm.nih.gov/bsd/authors1.html.

12. Additional benefits of multi-center trials unrelated to medical science should also be acknowledged. If a for-profit manufacturer is the trial's sponsor, involving more practitioners at more hospitals means more physicians prescribing a new drug or more surgeons preferring a new medical device are conveniently introduced to the new product sooner if the trial succeeds. Conversely, an insurance company may benefit from having more rather than fewer practitioners engaged in evaluating how a less expensive drug or procedure is equally or more efficacious and safe than the current approach. Furthermore, the pedigree of the practitioners and institutions performing the trial can be a helpful marketing element for convincing the rest of the medical community about the trials' legitimacy and the import of its conclusions.

13. There is a large and growing body of literature describing how pre-clinical results involving animal disease models and new medical countermeasures routinely fail to be reproduced when experiments are repeated by other laboratories, and what the research community is doing to rectify this situation. It is beyond the scope of this book to address this problem adequately so the reader will have to familiarize himself or herself with the subject if interested in learning more.

14. That is, unless the study was intended to seek regulatory approval for a novel veterinary drug, vaccine, or device under the auspices of the FDA's Center for Veterinary Medicine or its counterpart in another country. But such a purpose is unlikely, at least in the United States, since laboratory animals are so frequently administered a wide variety of medications under highly variable circumstances in the course of their use in biomedical research. That in turn means more professional judgment, and off-label use is often allowed for purely practical reasons.

15. Bhatt, D.L., Mehta, C., 2016. Adaptive designs for clinical trials. N. Engl. J. Med. 375, 65–74; Harrington, D., Parmigiani, G., 2016. I-SPY 2—a glimpse of the future of phase 2 drug development? N. Engl. J. Med. 375, 7–9.

16. Ervin, A.-M., Taylor, H.A., Ehrhardt, S., 2016. NIH policy on single-IRB review—a new era in multicenter studies. N. Engl. J. Med. 375 (24), 2315–2317.

17. Foltz, C.J., Fox, J.G., Yan, L., Shames, B., 1996. Evaluation of various oral antimicrobial formulations for eradication of *Helicobacter hepaticus*. Lab. Anim. Sci. 46 (2), 193–197; Marx, J.O., Vudathala, D., Murphy, L., Rankin, S., Hankenson, F.C., 2014. Antibiotic administration in the drinking water of mice. J. Am. Assoc. Lab. Anim. Sci. 53 (3), 301–306; Slate, A.R., Bandyopadhyay, S., Francis, K.P., Papich, M.G., Karolewski, B., Hod, E.A., Prestia, K.A., 2014. Efficacy of enrofloxacin in a mouse model of sepsis. J. Am. Assoc. Lab. Anim. Sci. 53 (4), 381–386.

18. See Chapter 5 for pertinent references and more details.

19. Baker, S.W., Prestia, K.A., Karolewski, B., 2014. Using reduced personal protective equipment in an endemically infected mouse colony. J. Am. Assoc. Lab. Anim. Sci. 53 (3), 273–277.

20. Scholthof, K.-B.G., Adkins, A., Czosnek, H., Palukaitis, P., Jacquot, E., Hohn, T., Hohn, B., Saunders, K., Candresse, T., Ahlquist, P., Hemenway, C., Foster, G.D., 2011. Top 10 plant viruses in molecular plant pathology. Molec. Plant Pathol. 12 (9), 938–954.

21. Balique, F., Lecoq, H., Raoult, D., Colson, P., 2015. Can plant viruses cross the kingdom border and be pathogenic to humans? Viruses 7, 2074–2098.

22. Sansone, E.B., Fox, J.G., 1977. Potential chemical contamination in animal feeding studies: evaluation of wire and solid bottom caging systems and gelled feed. Lab. Anim. Sci. 27 (4), 457–465.
23. Perry, S., Levasseur, J., Chan, A., Shea, T.B., 2008. Dietary supplementation with S-adenosyl methionine was associated with protracted reduction of seizures in a line of transgenic mice. Comp. Med. 58 (6), 604–606; Tsao, F., Culver, B., Pierre, J., Shanmuganayagam, D., Patten, C., Meyer, K.C., 2012. Effect of prophylactic supplementation with grape polyphenolics on endotoxin-induced serum secretory phospholipase A2 activity in rats. Comp. Med. 62 (4), 271–278; Ziegler, T., Sosa, M., Peterson, L., Colman, R., 2013. Using snacks high in fat and protein to improve glucoregulatory function in adolescent male marmosets (*Callithrix jacchus*). J. Am. Assoc. Lab. Anim. Sci. 52 (6), 756–762.
24. Canadian Agency for Drugs and Technologies in Health, 2016. Pre-operative carbohydrate loading or hydration: a review of clinical and cost-effectiveness, and guidelines. CADTH Rapid Response Reports. Available at: https://www.cadth.ca/pre-operative-carbohydrate-loading-or-hydration-review-clinical-and-cost-effectiveness-and.
25. Bouritius, H., van Hoorn, D.C., Oosting, A., van Middelaar-Voskuilen, M.C., van Limpt, C.J., Lamb, K.J., van Leeuwen, P.A., Vriesema, A.J., van Norren, K., 2008. Carbohydrate supplementation before operation retains intestinal barrier function and lowers bacterial translocation in a rat model of major abdominal surgery. J. Parenter. Enteral. Nutr. 32, 247–253.
26. Diks, J., van Hoorn, D.E., Nijveldt, R.J., Boelens, P.G., Hofman, Z., Bouritius, H., van Norren, K., van Leeuwen, P.A., 2005. Preoperative fasting: an outdated concept? J. Parenter. Enteral. Nutr. 29, 298–304.
27. http://www.tecniplast.it/usermedia/en/2016/brochures/TP_DIGILAB.pdf.

## Chapter 9

1. This conundrum of logic cuts both ways. Animal protectionists argue that because animals can and do suffer just like us, it is inhumane to use them in research that creates a suffering state, intentionally or otherwise. At the same time (and sometimes in the same sentence), they argue that because laboratory animals aren't humans, any conclusions arising from animal models of human illnesses are worthless as well as immoral. However, if their second argument is restated with respect to animal models of human suffering, then it appears that, according to their first argument, animal suffering should indeed provide an excellent model for studying the human version because it's identical.
2. Recognition and Alleviation of Pain and Distress in Laboratory Animals, 1992. National Research Council, National Academics of Press, Washington. Available at: http://www.nap.edu/download.php?record_id=1542.
3. Recognition and Alleviation of Distress in Laboratory Animals, 2008. National Research Council, National Academies Press, Washington. Available at: http://www.nap.edu/download.php?record_id=11931#.
4. This is a gross simplification of the panel's conclusions and one is encouraged to read the entire report for a complete picture.
5. Phelps, E.A., LeDoux, J.E., 2005. Contributions of the amygdala to emotion processing: from animal models to human behavior. Neuron 48, 175–187.
6. Is laughter just a human thing? Radiolab, National Public Radio, February 2008. http://www.radiolab.org/story/91589-is-laughter-just-a-human-thing/; Panksepp, J., Burgdorf, J., 2000. 50-kHz chirping (laughter?) in response to conditioned and unconditioned tickle-induced reward in rats: effects of social housing and genetic variables. Behav. Brain. Res. 115, 25–38.
7. Panksepp, J., 1998. Affective Neuroscience: The Foundations of Human and Animal Emotions. Oxford University Press, New York.

8. To the lay reader who regards animal emotions as obviously human-like, if not human, my epiphany may sound foolish and long overdue. But regardless of how apparent the similarity in emotional states between species may appear, one must be careful not to overinterpret one's observations through a single (and solely human) prism. A fundamental lesson I took away from veterinary school is to try to understand what's in the patient's mind when diagnosing a problem and administering treatment; thinking more like the dog, horse, or mouse under one's care is just as important as the therapy itself and can avoid making things worse. Just as important, medical progress is built on scientific evidence rather than testimonials, and an accumulation of objective data over time is the only means by which advances in knowledge are likely to be sustained.

9. Gottlieb, D.H., Capitanio, J.P., McCowan, B., 2013. Risk factors for stereotypic behavior and self-biting in rhesus macaques (Macaca mulatta): animal's history, current environment, and personality. Am. J. Primatol. 75 (10), 995–1008; Lutz, C.K., Williams, P.C., Sharp, R.M., 2014. Abnormal behavior and associated risk factors in captive baboons (*Papio hamadryas* spp.). Am. J. Primatol. 76 (4), 355–361.

10. Alpert, M., 2005. Every breath you take. Sci. Am. 292 (2), 94–96.

11. Perhaps we could have just as easily generated similar observations by using human observers or infrared cameras during periods of darkness in the animal rooms. But the presence of human observers would possibly alter how normal monkeys behaved and skewed our findings, while 24-h cameras can not integrate motion with physiological data as seamlessly as the vests. Another option for measuring stereotypical behavior was surgical implantation of various sensors connected to data capture equipment either via wireless telemetry or through cables running from the animal to a computer via a tether. But the invasive nature of that approach conflicted with the experiments to which these animals were assigned. Today, simpler alternatives than this vest or any of the modalities mentioned above are readily available, such as popular fitness tracking wearables. These can be inserted in a vest or embedded in a collar worn by monkeys and wirelessly synched to a smart phone app for easy data transfer. At least one laboratory of which I'm aware is employing these devices to monitor normal rhesus and squirrel monkeys non-invasively in neurobehavioral studies, and I've been told another laboratory is using them on macaques with stereotypy and self-injurious behavior.

12. Excerpted from Camacho, J., Ostertag, K., Derchak, A., Jarrell, D., Niemi, S., October 15–19, 2006. A novel means to characterize and monitor stereotypic behavior in macaques. Abstract #P22, 57th Annual Session. American Association for Laboratory Animal Science, Salt Lake City, UT (J. Amer. Assoc. Lab. Anim. Sci. 45 (4), 96, 2006).

13. Pritchett-Corning, K.R., Keefe, R., Garner, J.P., Gaskill, B.N., 2013. Can seeds help mice with the daily grind? Lab. Anim. 47 (4), 312–315.

14. Garner, J.P., Thogerson, C.M., Dufour, B.D., Würbel, H., Murray, J.D., Mench, J.A., 2011. Reverse-translational biomarker validation of abnormal repetitive behaviors in mice: an illustration of the 4P's modeling approach. Behav. Brain Res. 219 (2), 189–196.

15. Novak, M.A., Kinsey, J.H., Jorgensen, M.J., Hazen, T.J., 1998. Effects of puzzle feeders on pathological behavior in individually housed rhesus monkeys. Am. J. Primatol. 46 (3), 213–227; Fontenot, M.B., Wilkes, M.N., Lynch, C.S., 2006. Effects of outdoor housing on self-injurious and stereotypic behavior in adult male rhesus macaques (*Macaca mulatta*). J. Am. Assoc. Lab. Anim. Sci. 45 (5), 35–43.

16. Lutz, C.K., 2014. Stereotypic behavior in nonhuman primates as a model for the human condition. ILAR J. 55 (2), 284–296.

17. Russoniello, C.V., Fish, M., O'Brien, K., 2013. The efficacy of casual videogame play in reducing clinical depression: a randomized controlled study. Games Health J. 2 (6), 341–346; Fish, M.T., Russoniello, C.V., O'Brien, K., 2014. The efficacy of prescribed casual videogame play in reducing symptoms of anxiety: a randomized controlled study. Games Health J. 3 (5), 291–295.

18. This is merely another example of how man could serve as a "model" for animal medicine rather than the other way around as conventionally practiced, including for psychiatric disorders. Practicing cross-species modeling in the "opposite" direction is a tactic not employed often enough for this and other disease categories, but rich with promise.

19. Campos, M., Camacho, J.N., Duffy, A.M., Gale, J., Niemi, S.M., Dougherty, D.D., Eskandar, E., 2010. PET imaging of rhesus monkey with OCD-like behaviors. Annual Meeting, Society for Neuroscience Annual Meeting, Poster Presentation 2010-S-15348-SfN; Tien, A., Pearlson, G., Machlin, S., Bylsma, F., Hoehn-Saric, R., 1992. Oculomotor performance in obsessive compulsive disorder. Am. J. Psychiatry 149 (5), 641–646.

20. C.H.O.I.C.E. – The Smart Caging Choice. http://www.britzco.com/CatalogItem.aspx?Product Id=102300.

21. Camacho, J., Perlman, J., Sorrells, A., Britz, W., October 12, 2010. Interactive enrichment housing for non-human primates. American Association for Laboratory Animal Science 61st National Meeting, Atlanta, GA. Abstract # PS46. Available at: http://www.britzco.com/CustomDesign/PrefCage/Preference%20Cage.pdf.

22. Animal Welfare Information Center, National Agricultural Library, United States Department of Agriculture, 1985. Public Law 99-198, Food Security Act of 1985, Subtitle F – Animal Welfare. https://www.nal.usda.gov/awic/public-law-99-198-food-security-act-1985-subtitle-f-animal-welfare.

23. DiVincenti, L., Wyatt, J.D., 2011. Pair housing of macaques in research facilities: A science-based review of benefits and risks. J. Am. Assoc. Lab. Anim. Sci. 50 (6), 856–863.

24. A survey conducted more than 14 years ago at a representative primate research center, at a time when individually housed monkeys were still the norm across the country, showed that 89% (321/362) of rhesus maintained without a cage mate had behavioral abnormalities at some point in their captivity, ranging from mild stereotypy (pacing) to SIB, and more commonly occurring in adult males (see Lutz, C., Well, A., Novak, M., 2003. Stereotypic and self-injurious behavior in rhesus macaques: a survey and retrospective analysis of environment and early experience. Am. J. Primatol. 60 (1), 1–15).

25. Filby, A.L., Paull, G.C., Bartlett, E.J., Van Look, K.J.W., Tyler, C.R., 2010. Physiological and health consequences of social status in zebrafish (*Danio rerio*). Physiol. Behav. 101, 576–587.

26. Vitalo, A., Fricchione, J., Casali, M., Berdichevsky, Y., Hoge, E.A., Rauch, S.L., Berthiaume, F., Yarmush, M.L., Benson, H., Fricchione, G.L., Levine, J.B., 2009. Nest making and oxytocin comparably promote wound healing in isolation reared rats. PLoS One 4 (5), e5523.

27. Vitaloa, A.G., Gorantla, S., Fricchione, J.G., Scichilone, J.M., Camacho, J., Niemi, S.M., Denninger, J.W., Benson, H., Yarmush, M.L., Levine, J.B., 2012. Environmental enrichment with nesting material accelerates wound healing in isolation-reared rats. Behav. Brain Res. 226, 606–612.

28. van Dellen, A., Cordery, P.M., Spires, T.L., Blakemore, C., Hannan, A.J., 2008. Wheel running from a juvenile age delays onset of specific motor deficits but does not alter protein aggregate density in a mouse model of Huntington's disease. BMC Neurosci. 9:34–46; Wood, N.I., Carta, V., Milde, S., Skillings, E.A., McAllister, C.J., et al., 2010. Responses to environmental enrichment differ with sex and genotype in a transgenic mouse model of Huntington's Disease. PLoS One 5 (2), e9077. https://doi.org/10.1371/journal.pone.0009077.

29. Cao, L., Liu, X., Lin, E.D., Wang, C., Choi, E.Y., Riban, V., Lin, B., During, M.J., 2010. Environmental and genetic activation of a brain-adipocyte BDNF/leptin axis causes cancer remission and inhibition. Cell 142, 52–64; Kappeler, L., Meaney, M.J., 2010. Enriching stress research. Cell 142, 15–17.

30. Pedersen, L., Idorn, M., Olofsson, G.H., Pedersen, B.K., thor Straten, P., Hojman, P., 2016. Voluntary running suppresses tumor growth through epinephrine- and IL-6-dependent NK cell mobilization and redistribution. Cell Metab. 23, 554–562.

31. Wafer, L.N., Jensen, V.B., Whitney, J.C., Gomez, T.H., Flores, R., Goodwin, B.S., 2016. Effects of environmental enrichment on the fertility and fecundity of zebrafish (*Danio rerio*). J. Am. Assoc. Lab. Anim. Sci. 55 (3), 291–294.

32. Kaiser, T., Feng, G., 2015. Modeling psychiatric disorders for developing effective treatments. Nat. Med. 21 (9), 979–988.

33. Rogers, J., Raveendran, M., Fawcett, G.L., Fox, A.S., Shelton, S.E., Oler, J.A., Cheverud, J., Muzny, D.M., Gibbs, R.A., Davidson, R.J., Kalin, N.H., 2013. CRHR1 genotypes, neural circuits and the diathesis for anxiety and depression. Molec. Psychiatry 18 (6), 700–707; Tang, R., Noh, H.J., Wang, D., Sigurdsson, S., Swofford, R., Perloski, M., Duxbury, M., Patterson, E.E., Albright, J., Castelhano, M., Auton, A., Boyko, A.R., Feng, G., Lindblad-Toh, K., Karlsson, E.K., 2014. Candidate genes and functional noncoding variants identified in a canine model of obsessive-compulsive disorder. Genome Biol. 15 (3), R25–R39; Dodman, N.H., Ginns, E.I., Shuster, L., Moon- Fanelli, A.A., Galdzicka, M., Zheng, J., Ruhe, A.L., Neff, M.A., 2016. Genomic risk for severe canine compulsive disorder, a dog model of human OCD. Int. J. Appl. Res. Vet. Med. 14 (1), 1–18.

34. Pajer, K., Andrus, B.M., Gardner, W., Lourie, A., Strange, B., Campo, J., Bridge, J., Blizinsky, K., Dennis, K., Vedell, P., Churchill, G.A., Redei, E.E., 2012. Discovery of blood transcriptomic markers for depression in animal models and pilot validation in subjects with early-onset major depression. Transl. Psychiatry 2, e101. https://doi.org/10.1038/tp.2012.26.

35. Hoeft, F., McCandliss, B.D., Black, J.M., Gantman, A., Zakerani, N., Hulme, C., Lyytinen, H., Whitfield-Gabrieli, S., Glover, G.H., Reiss, A.L., Gabrieli, J.D.E., 2011. Neural systems predicting long-term outcome in dyslexia. Proc. Natl. Acad. Sci. U.S.A. 108 (1), 361–366.

36. Gregg, V.R., 2000. Parkinson's progress. Robert W. Woodruff Health Sciences Center of Emory University. http://www.whsc.emory.edu/_pubs/em/2000spring/parkinson.html; Deep brain stimulation for Parkinson's disease Part 1 and Part 2. https://www.youtube.com/watch?v=xejclvwbwsk and https://www.youtube.com/watch?v=IOHtUzW02cg.

37. Cleary, D.R., Ozpinar, A., Raslan, A.M., Ko, A.L., 2015. Deep brain stimulation for psychiatric disorders: where we are now. Neurosurg. Focus 38 (6), E2.

38. Bain, L., Posey Norris, S., Stroud, C., 2015. Non-Invasive Neuromodulation of the Central Nervous System. National Academies Press, Washington.

39. Braitman, L., 2014. Animal Madness: How Anxious Dogs, Compulsive Parrots, and Elephants in Recovery Help Us Understand Ourselves. Simon & Shuster, New York.

## Chapter 10

1. In case you're wondering, here are my other career guidelines, listed in no particular order: (1) hire people smarter than you to work under you and then get out of their way, after you've concluded they're of good character and enjoy working; (2) insist on excellence and never tolerate mediocrity—the minute you do, the entire organization knows it and your credibility is shot; (3) besides providing for my family, the most important component of any job for me is to continue learning—if I stop learning from the job I have, it's time to move on to something else; (4) I always want to be the first to anticipate my looming obsolescence, whether I'm still learning or not, so I can plan my next move with more forethought and less disruption. It's been a rewarding ride so far.

2. Around 75% occupancy on any given day was the threshold to break even on an operating basis, with 90%–95% occupancy needed to maximize net profitability. If you tried to run the business at 100% occupancy, you eliminated any cushion for unanticipated problems and risked exhausting your staff, as well.

3. Liker, J., 2004. The Toyota Way: 14 Management Principles from the World's Greatest Manufacturer. McGraw-Hill, New York.
4. NIH Data Book; NIH Budget History; NIH budget mechanism detail FY2001–14. http://report. nih.gov/NIHDatabook/Charts/Default.aspx?showm=Y&chartId=153&catId=1.
5. National Institutes of Health ARRA funding summary. https://report.nih.gov/recovery/NIH_ ARRA_Funding.pdf.
6. The MGH Center for Comparative Medicine remains a recognized leader in both lean vivarium management and participatory continuous improvement in our field, serving as an enlightened management model for other programs including the one I currently direct.
7. The reader is referred to my calculated savings in Chapter 5, Democratize the Guide, and to my presentation, "Sharing of Acceptable Performance Standards," at the ILAR Roundtable Workshop on Performance Standards, Washington, DC, April 21, 2015. http://nas-sites.org/ilar-roundtable/files/2015/05/NIEMI-ILAR-Performance-Standards-Workshop-Niemi-4.21.15.pdf. These figures and calculations are not proffered as conclusive but merely illustrative. Readers are encouraged to analyze their own institutions' research finance metrics as well as conduct broader studies for better grounded projections.
8. The latter justification is rightly decried by AAALAC spokespersons as "using the AAALAC club," i.e., using the threat of losing accreditation unless the institution agrees to funding the expense in question. What frustrates AAALAC leadership is that said threats are sometimes inaccurate, and more affordable solutions may be just as satisfactory for renewing accreditation for another 3 years before the next scheduled site visit.
9. I've encountered numerous directors in the non-profit sector who claim their program is entirely self-financing from per diem revenues. When pressed, they either didn't include or didn't know about all the indirect costs (e.g., utilities and building maintenance, depreciation and amortization, general and administration expenses at higher levels) that support their vivaria. These must be included in order to know the true financial cost of running a program, especially since the total prorated indirect expenses that are allocated to a program usually exceeds its direct expenses.
10. It was within such an environment that emphasized program outcome over process that Donna and I arrived at MGH. The hospital's research leadership did not direct us to instill a business culture or execute a novel organizational strategy for the program, and probably wouldn't have cared or known if we deviated from the usual model. Thus, we were on our own and, truth be told, we had to fabricate an atmosphere of urgency for change in order to move the needle. We were fortunate to work with a great crew that accepted the challenge and was game enough to try a different approach.
11. An alternative quick start for those who aren't familiar with accounting at all is Essentials of Accounting, 11th edition, by Breitner, L.K., Anthony, R.N., 2013. Pearson Education, Inc., New York. This is a self-paced primer that does a good job of introducing the subject.
12. National Center for Research Resources Office of Science Policy and Public Liaison, 2000. Cost Analysis and Rate Setting Manual for Animal Research Facilities, NIH Publication No. 00–2006, Bethesda. Available at: http://grants.nih.gov/grants/policy/air/rate_setting_manual_2000.pdf.
13. Survey respondents were asked to indicate their level of dissatisfaction for each bullet by selecting from the following list: "Never, Seldom, Occasionally, Often, Always, N/A."
14. Some may object that we didn't distinguish between "Occasionally" versus "Often" or "Always" dissatisfied when, in fact, there might be wide and potentially misleading disparities between these three opinions. But equating all three frequencies was intentional, so that occasional dissatisfaction was taken just as seriously as the two higher frequencies, as evidence

of our commitment to customer service. It also bears pointing out that this was not a regulatory compliance audit or accreditation site visit. That would have been a waste of time because it would have depended on how knowledgeable our researchers were about the AWA and the Guide. So we weren't asking researchers if they thought we were following the rules, but rather how they felt about how well we were meeting their needs. And because we emphasized dissatisfaction (we even capitalized the word in every survey question), the results were also good indicators of what our customers found pleasing or at least not irritating, such as the comportment of program staff and their rapport with animal users.

15. Bolfing, C.P., 1989. How do customers express dissatisfaction and what can service marketers do about it? J. Serv. Mark. 3 (2), 5–23. http://www.emeraldinsight.com/doi/pdfplus/10.1108/EUM0000000002483.

16. We invited users (and paid them again in per diem credits) to judge photographs of rodent cages at various stages of soiling and vote on when they thought a cage needed changing based on its appearance. The outcome of that exercise is included in Chapter 5, Democratize the Guide, Note 16.

17. Animal Welfare Regulations as of November 6, 2013 as found in the Code of Federal Regulations Title 9—Animals and Animal Products, Chapter 1—Animal and Plant Health Inspection Service, Department of Agriculture, Subchapter A—Animal Welfare, Parts 1–4, §2.31(d)(1)(vi): "The housing, feeding, and nonmedical care of the animals will be directed by the attending veterinarian *or other scientist* trained and experienced in the proper care, handling, and use of the species being maintained or studied" (emphasis added); The Guide, eighth ed. 2011, page 14: "The attending veterinarian (AV) is responsible for the health and wellbeing of all laboratory animals used at the institution... The AV should oversee other aspects of animal care and use (e.g., husbandry, housing) to ensure that the Program complies with the Guide." Neither directive prevents a non-veterinarian being appointed as a program Director who, in turn, oversees the program's head veterinarian.

## Chapter 11

1. Two excellent introductions to the lean management literature are Womack, J.P., Jones, D.T., 2003. Lean Thinking: Banish Waste and Create Wealth in Your Corporation, Revised and Updated. Free Press, New York; Goldratt, E.M., Cox, J., Whitford, D., 2004. The Goal: A Process of Ongoing Improvement, third ed., North River Press, Great Barrington.

2. While there is a voluminous bibliography about Toyota and its lean management philosophy and applications, I've found the following publications to be especially enlightening: Liker, J., 2004. The Toyota Way: 14 Management Principles from the World's Greatest Manufacturer. McGraw-Hill, New York; Mishina, K., Takeda, K., 1992. Toyota Motor Manufacturing, U.S.A., Inc. Harvard Business School Case 9-693-019, Revised 1995; Sobek, D.K., Smalley, A., 2008. Understanding A3 Thinking: A Critical Component of Toyota's PDCA Management System. Taylor & Francis Group, New York; Spear, S., Bowen, H.K., September–October, 1999. Decoding the DNA of the Toyota Production System. Harv. Bus. Rev. 96–106.

3. Spear, S., 2017. Fast discovery. The Health Foundation. Available at: http://www.health.org.uk/sites/health/files/FastDiscovery.pdf.

4. Wayland, M., 2015. Toyota's per-car profits lap Detroit's Big 3 automakers. Detroit News. Available at: www.detroitnews.com/story/business/autos/2015/02/22/toyota-per-car-profits-beat-ford-gm-chrysler/23852189/.

5. Oliva, R., Gittell, J.H., 2002. Southwest Airlines in Baltimore. Harvard Business School Case 9-602-156.

6. Renamed since as The Murli Group. www.The_Murli_Group.com.

7. Speaking of fear, we even got written up during IACUC semi-annual inspections while excessive inventory was being eliminated. IACUC inspectors were concerned we would run out of basic supplies, such as bedding or food, which could adversely impact animal welfare and research results, and included those concerns in their draft reports. They later became satisfied that our inventory monitoring process was reliable and that we could quickly procure such supplies on short notice if necessary, so those concerns were deleted from the final inspection report reviewed by the IACUC. However, I was more worried about a ridiculous precedent being established that would normalize conjectures as legitimate IACUC inspection findings. Luckily, we dodged that bullet.

8. Peter Drucker, my favorite and perhaps the greatest writer ever on the subject of management, stated it best: "There is only one valid definition of business purpose: *to create a customer...* It is the customer who determines what a business is. It is the customer alone whose willingness to pay for a good or for a service converts economic resources into wealth, things into goods. What the customer buys and considers value is never just a product. It is always a utility, that is, what the product or service does for him." From The Essential Drucker: The Best of 60 years of Peter Drucker's Essential Writings on Management. New York: Harper Business, 2001. One may argue that a program of laboratory animal care is "determined" by other stakeholders besides animal researchers. That's partially true if funding agencies are added to the mix because they're the ones actually paying for the services that programs provide, but the value for which they're paying is based on scientists' needs in animal experimentation, testing, or education; it's less true if one is referring to regulators whose role is to ensure that the rules are obeyed versus whatever value is created from that research.

9. A representative list of published lean-driven improvements in vivarium operations includes the following: Britz, W.R., May/June, 2004. Lean for lab animal managers. Anim. Lab. News; Kahn, N., Umrysh, B.M., 2008. Improving animal research facility operations through the application of lean principles. ILAR J. 49, e15–e22; Kelly, H., July/August, 2011. Lean in the lab: a primer. ALN World; Cosgrove, C., November/December, 2012. An overview of lean management in lab animal facilities. Anim. Lab. News; Bassuk, J.A., Washington, I.M., 2013. The A3 problem solving report: a 10-step scientific method to execute performance improvements in an academic research vivarium. PLoS One 8, e76833; Tummala, S., Granowski, J.A., 2014. Lean concepts for vivarium operational excellence. Lab. Anim. Sci. Prof. 2 (1), 26–30; Bassuk, J.A., Washington, I.M., 2014. Iterative development of visual control systems in a research vivarium. PLoS One 9, e90076; Zynda, J.R., 2015. A shift in designing cage-washing operations. Lab. Anim. 44 (4), 146–149; Ertl, C., Kukami, N., 2016. Workflow-centric laboratory design. Anim. Lab. News Mag. 14 (3), 14–16.

10. This is different than applying a hazardous substance on the outside of the animal or delivering it in food or a water bottle, when some of the chemical can spill into the bedding and coat the inside of the cage; in these cases, additional precautions are indicated. And if the substance is either a live pathogen (and thereby can multiply after administration) or is radioactive (and thereby can remain a risk for days, months, or years), then no shortcuts are appropriate.

11. U.S. Department of Health and Human Services. Biosafety in Microbiological and Biomedical Laboratories, fifth ed. HHS Publication No. (CDC) 21–1112, Revised December 2009 – see pages 16–19. Available at: https://www.cdc.gov/biosafety/publications/bmbl5/bmbl.pdf.

12. NIH Guidelines for Research Involving Recombinant or Synthetic Nucleic Acid Molecules, April 2016 – see preface to Appendix B. Available at: http://osp.od.nih.gov/sites/default/files/resources/NIH_Guidelines.pdf.

13. Gawande, A., 2009. The Checklist Manifesto. Metropolitan, New York.

14. Anthes, E., 2015. The trouble with checklists. Nature 523, 516–518.

15. Statistical Quality Control Handbook, second ed. (1958; copyright renewed 1984). AT&T Technologies, Inc., Delmar, Charlotte, NC.
16. Anon., January 18, 2014. Creating a business: Testing, testing. The Economist.
17. A good account about how an embrace of lean must be total and sustained in order to be transformational is provided in Lancaster, J., 2017. The Work of Management. A Daily Path to Sustainable Improvement. Lean Enterprise Institute, Inc., Cambridge, MA. A good overview of six sigma versus lean is provided in Nave, D., 2002. How to compare six sigma, lean and the theory of constraints. March issue, pages 73–78.

## Chapter 12

1. The same evolution applies to front office staff. What used to be done with pen and paper now requires familiarity with various computer programs as well as the computers themselves. Tracking and sharing transactions pertaining to finance, personnel, protocols, animals, vendors, regulatory compliance, information technology, etc. is way more complex and voluminous than ever before. However, the underlying skills needed for these tasks are less specialized and more easily transferable to other industries than those required for laboratory animal care. So while program administrators are greatly appreciated and highly valuable, they aren't the focus of this chapter.
2. Similar posters are available at http://www.criver.com/products-services/basic-research/poster/rodent-health-conditions and possibly elsewhere.
3. Animal technicians are also instructed that if they have any question or hesitation about an animal's condition or if it doesn't match what's on the health concern poster, then they are to contact a veterinary technician or veterinarian immediately rather than generate an e-mail message from the library. All our animal technicians in my current program carry workplace-restricted smart phones for this and other purposes so it's simple to pose a question if the diagnosis isn't obvious or doesn't match the poster. Each active health case is also monitored during regular rounds by veterinary staff, comparing those special red cage cards to the animal's current condition and updating the card and treatment as appropriate.
4. Ingram, L., Pina, A., Ehr, I., 2007. Standardized email templates for rodent health alerts. Abstract #350023, American Association for Laboratory Animal Science 58th Annual Meeting, Charlotte, NC.
5. Colby, S.L., Ortman, J.M., 2014. The Baby Boom Cohort in the United States: 2012–60. U.S. Department of Commerce, Economics and Statistics Administration, U.S. Census Bureau, P25-1141. Available at: https://www.census.gov/prod/2014pubs/p25-1141.pdf.
6. Technician Certification, American Association for Laboratory Animal Science. https://www.aalas.org/certification#.WM22-We1vIU.
7. Chapter 74 Vocational Technical Education Program Directory, Office for Career/Vocational Technical Education, Massachusetts Department of Elementary & Secondary Education. http://www.doe.mass.edu/cte/programs/directory.html.
8. Technical High Schools Transfer Agreements, Massachusetts Community Colleges. http://www.masscc.org/articulation.
9. Mass*Net Non-Credit Training Courses, Massachusetts Community Colleges. http://www.masscc.org/content/massnet-non-credit-training-courses.
10. Cohn, D., Passel, J.S., 2015. Chapter 2: Immigration's impact on past and future U.S. population change, in Modern Immigration Wave Brings 59 Million to U.S., Driving Population Growth and Change Through 2065: Views of Immigration's Impact on U.S. Society Mixed. Pew Research Center, Washington DC. Available at: http://www.pewhispanic.org/2015/09/28/chapter-2-immigrations-impact-on-past-and-future-u-s-population-change/.

11. A changing America. The Kiplinger Letter. Washington, DC, vol. 93, No. 50. December 16, 2016. Available at: http://www.kiplinger.com/pdf/kwl/index.php?file=kwl161216-2.html. (NB: this emphasis on Hispanics should not be considered a slight against any other demographic group in the United States. It's merely intended to reflect changing US demographics from which larger vs. smaller groups of new employees are likely to be found.)

12. Jarrell, D.M., Burleson, G.H., Niemi, S.M., 2005. A novel ESL training program for animal care staff. Abstract # PS22, 56th Annual Meeting, American Association for Laboratory Animal Science, Saint Louis, MO, November 6–9.

13. This emphasis on Spanish should in no way be interpreted as dismissive of other ethnic or racial groups in the United States. It's merely one approach to building more bridges to one of the largest communities of new employees likely to grow. Similar strategies can be easily envisioned to reach out to other peoples.

14. Global Partner Membership program. American Association for Laboratory Animal Science. https://www.aalas.org/membership/global-partner#.WM3Tjme1vIU.

# Chapter 13

1. Some institutions that are publicly funded, such as state universities and colleges, may be subject to particular sunshine laws that require public disclosure of AAALAC site visit details and correspondence. But these are the exception, not the rule.

2. Sometimes, not walking through every animal room during a site visit can be wrongly interpreted by the host's animal care staff as disinterest or a sign of disrespect on the part of the site visitors, especially after the staff have just spent the previous weeks or days making everything look excessively or suspiciously spiffy. That's why the opening remarks of any site visit should set expectations accordingly, stating that while the site visitors are very appreciative of the effort everyone has made in their preparations, please don't be disappointed if we don't go into every room; we're here to judge the program rather than conduct a white glove test on the facilities. If a program is well managed and follows the Guide, etc. as it's supposed to, an accreditation assessment could theoretically be conducted at any time, without much advance notice and avoiding all the scrambling that often takes place beforehand. The very first site visit in which I participated was at an institution where I was consulting as the Attending Veterinarian and therefore part of the host team. I remember an AAALAC Council member wryly remarking as we walked through the vivarium that this was the first time he had ever seen the new coat of paint still wet. Ever since, it's been my philosophy that whatever program I'm directing should be able to welcome inspectors and site visitors any and every day, offer to answer any questions they may have, and ask that they please don't get in the way as we perform our regular duties. In other words, no hand wringing, no fire drills, no drama.

3. https://grants.nih.gov/grants/olaw/sampledoc/cheklist.htm.

4. https://grants.nih.gov/grants/olaw/reporting_noncompliance.htm; https://grants.nih.gov/grants/guide/notice-files/NOT-OD-05-034.html; https://grants.nih.gov/grants/olaw/Departures_flow_chart.pdf; http://grants.nih.gov/grants/olaw/departures_table.pdf.

5. §2143(f)[1] and §2149(b), (c), (d) of the Animal Welfare Act, USC Title 7, Chapter 54. https://www.aphis.usda.gov/animal_welfare/downloads/Animal%20Care%20Blue%20Book%20-%202013%20-%20FINAL.pdf.

6. http://www.aaalac.org/accreditation/rules.cfm#hearings.

7. In cases where OLAW's veterinarians impose their professional judgment on an institution but where a legitimate difference of professional opinion could apply, I've wondered what would happen if an equal number or more of board-certified specialists in laboratory animal medicine submitted an amicus brief of sorts on behalf of that institution. What standing, especially amongst peer experts, would OLAW's judgment have then? Who would decide whose judgment is better qualified?.

8. We don't usually distinguish between good and great with respect to laboratory animal care. The only differentiation that's ever drawn to my knowledge is if your program has at least three consecutive AAALAC site visits with no mandatory corrections, then your subsequent site visits will be headed by former rather than current members of AAALAC Council, a small distinction to be sure. To be fair, I completely understand why AAALAC doesn't offer more than one level of accreditation. There's enough variety amongst their constituents to justify not making things even more complicated (and, therefore, even harder to compare programs).

9. Austin, R.D., Devin, L., Sullivan, E., July 7, 2008. Oops! Accidents lead to innovations. So, how do you create more accidents? Wall Str. J., R6.

10. It's highly likely that others' program management categories and priorities will differ from mine, but we'll go with this solely for the purpose of discussion.

11. Pyzdek, T., Keller, P., 2009. The Six Sigma Handbook, third ed. McGraw-Hill, New York.

12. Keller, P., 2011. Statistical Process Control Demystified. McGraw-Hill, New York.

13. Anon., March 26, 2016. How cities score. Better use of data could make cities more efficient—and more democratic. The Economist.

14. www.cityofboston.gov/cityscore/. Locals and baseball fans will appreciate how the posted "scoreboard" matches the look of the scoreboard on the famous "Green Monster" at Fenway Park, home of the Boston Red Sox.

15. The phrase, "in general," is not to be confused with the concern I voiced earlier about "close enough." The latter refers to an implicit resignation to occasional and sometimes catastrophic lapses, while the former is a management tool that recognizes the inevitability of errors but is used explicitly to reduce their occurrence and impact in routine vivarium activities.

16. A good analogy is the Dow Jones Industrial Average (DJIA), perhaps the most closely watched stock market index in the world today. Created in 1896 to track how the US manufacturing (industrial) sector was performing, it has since evolved to include companies from diverse sectors such as health care, banking, communications, insurance, food service, and entertainment. Moreover, the list of specific firms that comprise the DJIA isn't fixed but can change over time as individual firms' fortunes rise and fall and if some of the sectors in the DJIA are thought to be either over- or under-represented (for example, Apple replaced AT&T in 2015). The same philosophy can be applied to specific quality determinants of one's VivariumValue.

17. One may wish to consider an optional dimension in calculating a total VivariumValue. Before CityScore could become a reality, there had to be extensive discussions about what to include and how to assign relative significance to the various inputs. Are potholes as important as stabbings or power failures? Will food poisoning outbreaks be given the same emphasis as highway toll collections or mass transit ridership? Similarly, would it be more informative to assign different weights to the various quality parameters employed for laboratory animal care and use? For example, 40% of the VivariumValue could be provided by metrics on the basis of human health and safety, 40% on animal welfare, 15% on institutional behavior, and 5% on community and public engagement. Alternatively, every input could have an equal arithmetic representation in one's VivariumValue. Either approach has merit, requiring yet another decision for program leadership to make.

# Chapter 14

1. Ford, H., 1922. My Life and Work. Doubleday, Page & Co., Garden City, NY.

2. Gross, D., 1996. Greatest Business Stories of All Time. John Wiley & Sons, Inc., New York, NY.

3. This generalization doesn't apply to situations, such as injecting rodents with small doses of radioactive tracers, where it's appropriate for researchers rather than animal care staff to perform routine husbandry tasks. In this case, there's no reason to train animal technicians on

radiation safety practices and assign them radiation exposure badges for only a small number of cages while research staff in that laboratory are already knowledgeable about handling isotopes and enrolled in radiation safety monitoring programs. But even while the researchers themselves are handling and changing rodent cages, they still usually have to pay the regular per diem price.

4. As the magnitude of irreproducible published data involving animal research was becoming evident several years ago, there were earnest discussions in my field to consider establishing standardized (i.e., inviolate) husbandry arrays for various species. The reasoning was that if literally every program used exactly the same type of everything for their animals, data variability would decrease and data reproducibility would return. That reasoning perhaps was applicable to animal toxicity testing protocols established for regulatory submission. But it's entirely the wrong approach for discovery and translational research since all flexibility inherent in scientific pursuits would have been eliminated, and many institutions would have to replace their capital equipment inventory if they weren't lucky enough to possess the winning package. And even pre-clinical (animal) product testing needs some wiggle room to advance the 3R's. Thankfully, these conversations quietly dissolved, sparing the biomedical research community from an intellectual straightjacket that it couldn't afford anyway.

5. Hunt, P.A., et al., 2003. Bisphenol A exposure causes meiotic aneuploidy in the female mouse. Curr. Biol. 13 (7), 546–553.

6. Leys, L.J., McGaraughty, S., Radek, R.J., 2012. Rats housed on corncob bedding show less slow-wave sleep. J. Am. Assoc. Lab. Anim. Sci. 51 (6), 764–768; Robinson-Junker, A., Morin, A., Pritchett-Corning, K., Gaskill, B.N., 2017. Sorting it out: bedding particle size and nesting material processing method affect nest complexity. Lab. Anim. (UK) 51 (2), 170–180.

7. Carbone, E.T., Kass, P.H., Evan, K.D., 2016. Feasibility of using rice hulls as bedding for laboratory mice. J. Am. Assoc. Lab. Anim. Sci. 55 (3), 268–276.

8. Markets of One: Creating Customer-Unique Value through Mass Customization. Gilmore, J.H., Pine, B.J. (Eds.), 2000. Harvard Business School Publishing, Boston.

9. There are plenty of cheap and user-friendly smartphone apps one could adapt for constructing husbandry menus, and it's likely that the variety of apps will only increase in the future while prices remain attractive. We looked at several apps and selected AppSheet (Seattle, WA; www.appsheet.com) for starters.

10. Adjacent to all of these items in the display case will be our standard combination for mice and standard combination for rats. No QR codes will be on those cages because they are the default selections so no (custom) order can be placed for them.

11. Iyengar, S., 2010. The Art of Choosing. Hatchette Book Group, Inc., New York; for a lighter version, see Iyengar, S., 2011. How to make choosing easier. http://www.ted.com/talks/sheena_iyengar_choosing_what_to_choose. For details of the jam exercise, see Iyengar, S.S., Lepper, M.E., 2000. When choice is demotivating: can one desire too much of a good thing? J. Personal. Soc. Psychol. 79 (6), 995–1006.

## Chapter 15

1. One could say the clock started in 1885 when Louis Pasteur's rabies vaccine saved Jacob Meister, a 9-year old boy mauled by a rabid dog, after Pasteur had experimented with the vaccine on 50 dogs. Other animal research certainly occurred earlier in Europe and elsewhere, evidenced by Jeremy Bentham voicing no objection in 1825 to performing painful experiments on dogs when such experiments are intended to benefit mankind and likely to succeed (even though, ironically, he is also considered a patron saint of animal rights by many). But Pasteur's life-saving research that consumed many dogs as well as rabbits is as good a starting point as any.

2. Leslie, M., 2016. Whatever happened to.... Science 353 (6305), 1198–1201.

3. Gallup Poll, 2016. Birth control, divorce top list of morally acceptable issues. http://www.gallup.com/poll/192404/birth-control-divorce-top-list-morally-acceptable-issues.aspx.

4. Riffkin, R., 2015. In U.S., more say animals should have same rights as people. http://www.gallup.com/poll/183275/say-animals-rights-people.aspx.

5. I've been fascinated for a long time about people's seemingly universal attraction to animals (at least in wealthier countries and perhaps with the exception of snakes). Shortly after arriving at MGH, I had the pleasure of helping to get its pet therapy program started. Owners were lining up to bring their pet dogs into the hospital so patients could see them and perhaps give them a friendly pat, and it's become a very popular program at the hospital and in many other health care settings. It was especially fun to observe what happened when pet therapy dogs were brought into the elevator to visit different wards—everyone immediately brightened up even if they were employees or just other visitors. I became intrigued about what was actually happening in human brains that connected us to animals in such a consistently positive manner. So I seized an opportunity to organize a clinical trial with human subjects at MGH to investigate this further. Our study recruited moms who had both young children and pet dogs, to capture and analyze their reactions to photos of cute kids (theirs and others) versus photos of cute dogs (theirs and others). We even performed brain scans to see how their emotional centers reacted when viewing those photos. We were able to recruit 14 volunteers who qualified for the study and found significant differences between responses of maternal attachment and pet owner attachment (see Stoeckel, L.E., Palley, L.S., Gollub, R.L., Niemi, S.M., Evins, A.E., 2014. Patterns of brain activation when mothers view their own child and dog: an fMRI study. PLoS One 9 (10), e107205).

6. September 3, 2016. Better and better. The Economist.

7. Starr, D., April 18, 2004. A dog's life. Boston Globe Sunday Magazine.

8. The current crisis of experimental reproducibility, including but not limited to animal research, deserves more space than can be given adequate coverage in this book. The reader is encouraged to search the web for the many relevant news articles and commentaries on this topic, as well as review the proceedings of a workshop convened by ILAR in 2014. https://www.nap.edu/catalog/21835/reproducibility-issues-in-research-with-animals-and-animal-models-workshop.

9. Kirschner, M., 2013. A perverted view of "impact." Science 340 (6138), 1265.

10. Angus, D.C., van der Poll, T., 2013. Severe sepsis and septic shock. N. Engl. J. Med. 369 (9), 840–851.

11. Cohen, J., 2016. Surprising treatment "cures" monkey HIV infection. Science 354 (6309), 157–158.

12. Sharpe, A.H., Hunter, J.J., Ruprecht, R.M., Jaenisch, R., 1988. Maternal transmission of retroviral disease: transgenic mice as a rapid test system for evaluating perinatal and transplacental antiretroviral therapy. Proc. Natl. Acad. Sci. U.S.A. 85 (24), 9792–9796.

13. As is the case with other highly regulated industries that depend on independent audits and official inspections to verify compliance, if it wasn't documented, then it didn't happen. So if a required activity, such as recording that post-operative analgesics were administered at a date and time required in the protocol, is not recorded correctly or omitted, it is a justifiable violation even if it actually occurred.

14. These results and calculations are not intended to represent all vivaria, and may differ substantially from other programs for a variety of reasons. I encourage others to compile their own data to determine if there are opportunities for improvement locally, regardless of how their numbers compare to the data and conclusions presented here.

15. Fletcher, L., Ghadishah, A., 2009. Ex-Employees Claim "Horrific" Treatment of Primates at Lab. ABC News Nightline. http://abcnews.go.com/Nightline/story?id=6997869&page=1.
16. I once asked a distinguished senior representative of an animal protectionist group who sometimes speaks at animal research conferences why aren't we similarly invited to meetings of the other side? I was told that we probably wouldn't be treated as courteously (albeit with distrust) as he usually is at our side's meetings.
17. Fowler, A., Gamble, N.N., Hogan, F.X., Kogut, M., McCormick, M., Thorp, B., January 28, 2001. Talking with the enemy. Boston Globe.

## Chapter 16

1. Russell, W.M.S., Burch, R.L., 1959. The Principles of Humane Experimental Technique, Charles C. Thomas, publisher. Available on-line at: http://altweb.jhsph.edu/pubs/books/humane_exp/het-toc.
2. §2143 (a)(3)(B) and (7)(B)(i), Animal Welfare Act; §2.31(d)(1)(ii) and (iii), §2.32(c)(5)(ii) and (iii), §2.36(b)(2), Animal Welfare Regulations. U.S. Department of Agriculture Animal and Plant Health Inspection Service.
3. §495(c),(1)(B), U.S. Health Research Extension Act of 1985, Public Law 99-158, November 20, 1985, "Animals in Research"; Parts III and IV, U.S. Government Principles for the Utilization and Care of Vertebrate Animals Used in Testing, Research, and Training; Part IV.A.1.g., C.1.a. and b., D.1.b. and d. Public Health Service Policy on Humane Care and Use of Laboratory Animals.
4. The Guide, pp. 4–5 and 25–26.
5. Macleod, M., June 4–5, 2014. Restoring faith in the research enterprise – a call to action. Presented at the ILAR Roundtable Workshop on Reproducibility, Washington, DC. Available at: http://nas-sites.org/ilar-roundtable/roundtable-activities/reproducibility/webcast/panel-1-macleod-bourne/).
6. Freedman, L.P., Cockburn, I.M., Simcoe, T.S., 2015. The economics of reproducibility in preclinical research. PLoS Biol. 13 (6), e1002165.
7. Begley, C.G., 2014. Six red flags for suspect work. Nature 497, 433–434.
8. Landis, S.C., et al., (35 authors total) 2012. A call for transparent reporting to optimize the predictive value of preclinical research. Nature 490, 187–191.
9. Reference #1 above—Chapter 6. Reduction.
10. Tannenbaum, J., Bennett, B.T., 2015. Russell and Burch's 3Rs then and now: the need for clarity in definition and purpose. J. Am. Assoc. Lab. Anim. Sci. 54 (2), 120–132.
11. For example: Siesler, H.W. (Ed.), 2012. Biomedical Imaging: Principles and Applications. John Wiley & Sons, Hoboken, NJ; Eferl, R., Casanova, E. (Eds.), 2015, Mouse Models of Cancer: Methods and Protocols. Springer, New York.
12. For example: Barbosa, J.S., Sanchez-Gonzalez, R., Di Giaimo, R., Baumgart, E.V., Theis, F.J., Götz, M., Ninkovic, J., 2105. Live imaging of adult neural stem cell behavior in the intact and injured zebrafish brain. Science 348 (6236), 789–793.
13. Examples include the following guidelines:
    • US FDA Toxicological Principles for the Safety Assessment of Food Ingredients. Redbook 2000. Chapter IV.C.4.a. Subchronic Toxicity Studies with Rodents, Section II.D. (2003): "In general, for subchronic toxicity studies, experimental and control groups should have at least 20 rodents per sex per group. Ten rodents/sex/group may be acceptable for subchronic rodent studies when the study is considered to be range-finding in nature or when longer term studies are anticipated. These recommendations will help ensure that the number of animals that

survive until the end of the study will be sufficient to permit a meaningful evaluation of toxicological effects. If interim necropsies are planned, the number of animals per sex per group should be increased by the number scheduled to be sacrificed before completion of the study; for rodents, at least 10 animals per sex per group should be available for interim necropsy."

- US FDA Toxicological Principles for the Safety Assessment of Food Ingredients. Redbook 2000. Chapter IV.C.4.b. Subchronic Toxicity Studies with Non-Rodents, Section II.D. (2003): "In general, for subchronic toxicity studies, experimental and control groups should have at least 4 dogs per sex per group. These recommendations will help ensure that the number of animals that survive until the end of the study will be sufficient to permit a meaningful evaluation of toxicological effects. If interim necropsies are planned, the number of animals per sex per group should be increased by the number scheduled to be sacrificed before completion of the study."

- US EPA Health Effects Test Guidelines: OPPTS 870.3100 90-Day Oral Toxicity in Rodents [EPA 712–C–98–199](1998), Section (e)(iv)(A): "At least 20 rodents (10 males and 10 females) at each dose level."

- US EPA Health Effects Test Guidelines: OPPTS 870.3150 90-Day Oral Toxicity in Nonrodents [EPA 712–C–98–200] (1998), Section (e)(iv)(A): "At least eight animals (four females and four males) should be used at each dose level."

14. But regulators also advise on the risks of using too few animals: "The number of animals used per dose has a direct bearing on the ability to detect toxicity. A small sample size may lead to failure to observe toxic events due to observed frequency alone regardless of severity. The limitations that are imposed by sample size, as often is the case for NHP studies, may be in part compensated by increasing the frequency and duration of monitoring" (US FDA, July 1997. Guidance for Industry S6 Preclinical Safety Evaluation of Biotechnology-Derived Pharmaceuticals. Section 3.4, Number/ Gender of Animals).

15. The discussion about Replacing animal testing with non-animal alternatives is saved for Chapter 19.

16. ILAR, 2008. Recognition and Alleviation of Distress in Laboratory Animals. National Academy Press, Washington DC (see Chapter 9).

17. Precision medicine is not a concept applicable just to human patients. Veterinary students learn early in their studies about various health problems to which each breed of dog and cat is predisposed, knowledge used to practice precision veterinary medicine at the breed level long before the phrase had been coined. As we learn more about the genetic basis of each of these predispositions and discover which drugs are compatible or not with each genotype's metabolic properties, it's easy to imagine a time soon when breed-specific therapies will be *de rigueur*.

18. Schork, N.J., 2015. Time for one-person trials. Nature 520, 609–611.

19. Ferster, C.B., Skinner, B.F., 1957. Schedules of Reinforcement. Appleton-Century-Crofts, New York.

20. Kazdin, A.E., 2011. Single-Case Research Designs: Methods for Clinical and Applied Settings, second ed. Oxford University Press, New York.

21. U.S. Food and Drug Administration Center for Drug Evaluation and Research, August 1996. Guidance for industry. Single dose acute toxicity testing for pharmaceuticals.

22. Stokes, W.S., 2014. Validation and regulatory acceptance of toxicological testing methods and strategies. In Principles and Methods of Toxicology, sixth edition, Hayes, A.W., Kruger, C.L. (Eds.), CRC Press, Boca Raton.

23. Botham, P.A., 2002. Acute systemic toxicity. ILAR J. 43 (S), S27–S30.

24. Reference #1 above - Chapter 5. Replacement.

25. And perhaps free-living metazoans, as mulled over but not resolved by the authors.
26. www.theonehealthcompany.com.

## Chapter 17

1. A good example is a recent report issued by a trans-Atlantic working group of laboratory animal veterinarians tasked with evaluating the concepts of "harm" and "benefit" to both animals and humans on whose behalf these animals are used. Their assignment was to provide investigators, IACUCs, regulators, and the public in the United States and Europe a systematic way to perform a harm–benefit analysis (see citations below). This report was presented at a national IACUC conference shortly after its release and received a mixed reception from those in the breakout discussion. One criticism voiced was that the report appeared to accentuate (animal) harm while minimizing (human) benefit, thereby discounting or discrediting the contributions of animal-based research and testing. But such concerns aren't unique to this report. Any harm–benefit calculus involving laboratory animals can only assess obvious or presumed adverse effects on animal subjects (the harm), but has no way of accurately predicting the worth of an experiment or assay until it's completed (the benefit) and perhaps not evident until years later. And why doesn't the definition of "harm" include the real pain or distress that human patients continue to experience while waiting for a better medical countermeasure. Either way, how can one compare knowns to unknowns consistently and fairly? Either way, how can one ever compare knowns, such as animal harm, to unknowns, such as human benefit", consistently and fairly?
   Brønstad, A., Newcomer, C.E., Decelle, T., Everitt, J.I., Guillen, J., Laber, K., 2016. Current concepts of harm–benefit analysis of animal experiments—Report from the AALAS–FELASA Working Group on Harm–Benefit Analysis, part 1. Lab. Anim. (UK) 50 (1S), 1–20; Laber, K., Newcomer, C.E., Decelle, T., Everitt, J.I., Guillen, J., Brønstad, A., 2016. Recommendations for addressing harm–benefit analysis and implementation in ethical evaluation—Report from the AALAS–FELASA Working Group on Harm–Benefit Analysis, part 2. Lab. Anim. (UK) 50 (1S), 21–42.
2. § 2143(a)(3), Animal Welfare Act, and Animal Welfare Regulations, as of January 1, 2017. United States Department of Agriculture Animal and Plant Health Inspection Service publication APHIS 41-35-076; § 495(a)(2)(A and B) and (c)(1)(B), Health Research Extension Act of 1985, Public Law 99-158, November 20, 1985, "Animals in Research." Public Health Service Policy on Humane Care and Use of Laboratory Animals, NIH Publication No. 15-8013.
3. § 2143(a)(6)(A), Animal Welfare Act, and Part 2, Subpart C, § 2.31(a), Animal Welfare Regulations.
4. § 495(a)(2), Health Research Extension Act of 1985, Public Law 99-158, November 20, 1985, "Animals in Research."
5. Spielmann, H., 2002. Animal use in the safety evaluation of chemicals: harmonization and emerging needs. ILAR J. 43, S11–S17.
6. Occasionally I hear or read someone citing FDA requirements as justification for lethal endpoints (i.e., not permitting euthanasia intervention) in protocols intended for agency review that involve pre-clinical toxicity or efficacy testing. In these assays, it's important that some animals should get very sick for the purpose of establishing upper safety limits of a new drug for eventual clinical trials approval. If the test animals don't get sick enough, there is legitimate concern that the dose wasn't high enough so the test may have to be repeated. Similarly, EPA requirements are sometimes misrepresented to rationalize lethal endpoints when a company must establish the safety profile of an industrial or household chemical by animal testing. But both agencies permit moribund euthanasia so these stances are not valid.

7. My reasoning, by contrast, agrees with others who apply Category E if any time elapses after the animal is first observed to be in pain or distress and no prompt relief is allowed. This is not to say Category E procedures must be avoided even if scientifically justified, only that the Category definitions should be applied honestly. For example, if one is studying pain, no matter how mild, analgesics are contraindicated because these drugs would compromise the results. If one is studying behavioral distress, even without long-term effects, then the progression of that distress should not be impeded. In both cases, Category E is the appropriate designation because the animals were permitted to continue to experience those insults without immediate amelioration.

8. Mildvan, D., 2000. Surrogate markers: the AIDS clinical trials. Arch. Neurol. 57 (8), 1233–1234.

9. Wolfensohn, S., Hawkins, P., Lilley, E., Anthony, D., Chambers, C., Lane, S., Lawton, M., Voipio, H.-M., Woodhall, G., 2013. Reducing suffering in experimental autoimmune encephalomyelitis (EAE). J. Pharmacol. Toxicol. Methods 67, 169–176.

10. Reiser, J., von Gersdorff, G., Loos, M., Oh, J., Asanuma, K., Giardino, L., Rastaldi, M.P., Calvaresi, N., Watanabe, H., Schwarz, K., Faul, C., Kretzler, M., Davidson, M., Sugimoto, H., Kalluri, R., Sharpe, A.H., Kreidberg, J.A., Mundel, P., 2004. Induction of B7-1 in podocytes is associated with nephrotic syndrome. J. Clin. Investig. 113, 1390–1397 (NB: this paper suggests that proteinuria can be an early marker for nephrosis induced by endotoxin); Editorial, 2010. Biomarkers on a roll. Nat. Biotech. 28 (5), 431.

11. Morton, D.B., 2000. A systematic approach for establishing humane endpoints. ILAR J. 41 (2), 80–86.

12. Sotocinal, S.G., Sorge, R.E., Zaloum, A., Tuttle, A.H., Martin, L.J., Wieskopf, J.S., Mapplebeck, J.C.S., Wei, P., Zhan, S., Zhang, S., McDougall, J.J., King, O.D., Mogil, J.S., 2011. The Rat Grimace Scale: a partially automated method for quantifying pain in the laboratory rat via facial expressions. Molec. Pain 7, 55; Gaskill, B.N., Pritchett-Corning, K.R., 2016. Nest building as an indicator of illness in laboratory mice. Appl. Anim. Behav. Sci. https://doi.org/10.1016/j.applanim.2016.04.008; Dunbar, M. L., David, E. M., Aline, M. R., Lofgren, J. L., 2016. Validation of a behavioral ethogram for assessing postoperative pain in guinea pigs (*Cavia porcellus*). J. Am. Assoc. Lab. Anim. Sci. 55 (1), 29–34.

13. Martín, D., 2010. Functions of nuclear receptors in insect development, in Nuclear Receptors: Current Concepts and Future Challenges, Bunce, C.M., Campbell, C.J. (Eds.), Springer, New York.

14. Panksepp and Lahvis reviewed studies that indicated empathy between paired rodent conspecifics, one in pain and the other not (Jules, B., Panksepp, J.P., Lahvis, G.P., 2011. Rodent empathy and affective neuroscience. Neurosci. Biobehav. Rev. 35, 1864–1875). Does it follow that the animal in pain could derive comfort from its cage mate (Smith, M.L., Hostetler, C.M., Heinricher, M.M., Ryabinin, A.E., 2016. Social transfer of pain in mice. Sci. Adv. 2, e1600855)?.

15. Niemi, S.M., 2013. Laboratory animals as veterinary patients. J. Am. Vet. Med. Assoc. 242 (8), 1063–1065.

16. Knaus, W.A., Wagner, D.P., Draper, E.A., Zimmerman, J.E., Bergner, M., Bastos, P.G., Sirio, C.A., Murphy, D.J., Lotring, T., Damiano, A., 1991. The APACHE III prognostic system. Risk prediction of hospital mortality for critically ill hospitalized adults. Chest 100 (6), 1619–1636.

17. Zimmerman, J.E., Kramer, A.A., McNair, D.S., Malila, F.M., 2006. Acute Physiology and Chronic Health Evaluation (APACHE) IV: hospital mortality assessment for today's critically ill patients. Crit. Care Med. 34 (5), 1297–1310.

18. http://intensivecarenetwork.com/Calculators/Files/Apache4.html.

19. Goggs, R.A.N., Lewis, D.H.L. Multiple organ dysfunction syndrome (Chapter 7); Hayes, G., Matthews, K.A. Illness severity scores in veterinary medicine (Chapter 13). In Small Animal Critical Care Medicine—E-Book, Silverstein, D., Hopper, K. (Eds.), 2014. Elsevier Health Sciences.

20. Parra, N.C., Ege, C.A., Ledney, G.D., 2007. Retrospective analyses of serum lipids and lipoproteins and severity of disease in $^{60}$Co-irradiated *Sus scrofa domestica* and *Macaca mulatta*. Comp. Med. 57 (3), 298–304; Faix, J.D., 2013. Biomarkers of sepsis. Crit. Rev. Clin. Lab. Sci. 50 (1), 23–36; Shrum, B., Anantha, R.V., Xu, S.X., Donnelly, M., Haeryfar, S.M.M., McCormick, J.K., Mele, T., 2014. A robust scoring system to evaluate sepsis severity in an animal model. BMC Res. Notes 7, 233–244; Koch, A., Gulani, J., King, G., Hieber, K., Chappell, M., Ossetrova, N., 2016. Establishment of early endpoints in mouse total-body irradiation model. PLoS One 11 (8), e0161079.

21. Lambeth, S.P., Schapiro, S.J., Bernacky, B.J., Wilkerson, G.K., 2013. Establishing 'quality of life' parameters using behavioural guidelines for humane euthanasia of captive non-human primates. Anim. Welf. 22 (4), 429–435.

22. Goldberg, K., McDonald, C., Kiselow, M., 2014. Position Statements, Veterinary Society for Hospice and Palliative Care. http://www.vethospicesociety.org/position-statements/; American Veterinary Medical Association. Guidelines for Veterinary Hospice Care. https://www.avma.org/KB/Policies/Pages/Guidelines-for-Veterinary-Hospice-Care.aspx.

23. Kelly, A.S., Morrison, R.S., 2015. Palliative care for the seriously ill. N. Eng. J. Med. 373 (8), 747–755.

24. Official positions are mixed on this point:
    - The AWA states that "at least one member... shall not be affiliated in any way with such facility other than as a member of the Committee" but does not further prescribe that member's background except that such member "is intended to provide representation for general community interests in the proper care and treatment of animals," §2143(b)(1)(B). This undefined background is repeated in the accompanying Animal Welfare Regulations, Part 1, §1.1 Definitions ("Committee") and Part 2, Subpart C, §2.31(b)(3)(ii). By contrast, the current USDA Animal Care Policy Manual, dated May 23, 2016, includes Policy No. 15, Institutional Official and IACUC Membership, dated March 25, 2011, that states while "a veterinarian who is not the attending veterinarian may assume any one of the other program positions" (including Nonaffiliated Member), "APHIS has determined the nonaffiliated member should not be a laboratory animal user at any research facility." However, because USDA Animal Care Policies do not have regulatory standing and, therefore, cannot be enforced, there is theoretically some flexibility on this matter.
    - The Public Health Service Policy on Humane Care and Use of Laboratory Animals requires only "one individual who is not affiliated with the institution in any way other than as a member of the IACUC, and is not a member of the immediate family of a person who is affiliated with the institution" (IV.A.3.b.(4)) but does not further prescribe that individual's background. But while this PHS Policy allows that "An individual who meets the requirements of more than one of the categories detailed in IV.A.3.b.(1)-(4) of this Policy may fulfill more than one requirement" ((IV.A.3.c.), OLAW similarly doesn't prohibit a nonaffiliated member coming from the laboratory animal field (ARENA-OLAW Institutional Animal Care and Use Committee Guidebook, second edition, 2002, page 13), but "strongly recommends against the same person serving multiple roles because the responsibilities and authorities vested in each of the positions are distinct and often require different skills. Appointing one individual to more than one of these roles may circumvent intended checks and balances. Also of importance is the perception of conflict of interest, which can lead to allegations of improprieties from various sources." ARENA-OLAW Institutional Animal Care and Use Committee Guidebook, second edition, 2002, page 12).
    - According to the current Guide, "Public members should not be laboratory animal users" (page 24).

25. Begley, C.G., 2013. Six red flags for suspect work. Nature 497, 433–434.

26. Beresford, L., 2012. Teach back communication strategy helps healthcare providers help their patients. The Hospitalist 8. http://www.the-hospitalist.org/hospitalist/article/125116/teach-back-communication-strategy-helps-healthcare-providers-help-their; U.S. Department of Health & Human Services, 2015. Use the teach-back method: tool #5. Agency for Healthcare Research and Quality, Rockville, MD. http://www.ahrq.gov/professionals/quality-patient-safety/quality-resources/tools/literacy-toolkit/healthlittoolkit2-tool5.html.

27. Bankert, E., April 21, 2017. Informed Consent: Improving the Process and Interpreting the Principles in the Revised Regulations. Presentation at "The Three I's & Biosecurity: Creating Connections, Sharing Solutions & Building Strategies" conference organized by the Massachusetts Society for Medical Research, Providence RI.

28. I avoided including non-affiliated scientists as new IACUC members for two reasons. First, a laboratory animal veterinarian will have a broader knowledge base involving species biology, behavior, medicine, and welfare impacts for a wider variety of animal models. Second, laboratory animal veterinarians are more accustomed to being advocates for animal subjects, whereas scientists have additional job responsibilities, career goals, and demands on their time. So while scientists could certainly be effective "outsiders" in reviewing protocols, their contributions toward Refinement would likely be less consistent.

29. Bennett, BT., January 10, 2017. The 2016 USDA inspection data: what a difference a decade makes. National Association for Biomedical Research webinar.

30. Domestic Institutions with a PHS Approved Animal Welfare Assurance, 2016. Office of Laboratory Animal Welfare, National Institutes of Health. https://grants.nih.gov/grants/olaw/assurance/300index_name.htm.

31. Accredited Organizations—United States, AAALAC International. https://www.aaalac.org/accreditedorgsdirectorysearch/aaalacprgms.cfm.

32. A special issue of the journal, Theoretical Medicine and Bioethics, published seven essays along this line of thinking in 2014 after publication of the Institute of Medicine's 2011 report regarding the necessity of chimpanzees in biomedical research going forward and NIH's subsequent decision in 2013 to discontinue using the species. In that issue, see Kahn J. Lessons learned: Challenges in applying current constraints on research on chimpanzees to other animals. https://doi.org/10.1007/s11017-014-9284-6; Ferdowsian, H., Fuentes, A., Harms and deprivation of benefits for nonhuman primates used in research. https://doi.org/10.1007/s11017-014-9288-2; Wendler, D., Should protections for research with humans who cannot consent apply to research with nonhuman primates? https://doi.org/10.1007/s11017-014-9285-5.

33. DeMartino, E.S., Dudzinski, D.M., Doyle, C.K., Sperry, B.P., Gregory, S.E., Siegler, M., Sulmasy, D.P., Mueller, P.S., Kramer, D.B., 2017.Who decides when a patient can't? Statutes on alternate decision makers. N. Engl. J. Med. 376, 1478–1482.34.

34. Guidance for Industry 85, Good Clinical Practice, VICH GL9, June 8, 2011. US Food and Drug Administration, Center for Veterinary Medicine.

35. Kraft, R., Herndon, D.N., Finnerty, C.C., Shahrokhi, S., Jeschke, M.G., 2014. Occurrence of multi-organ dysfunction in pediatric burn patients - incidence and clinical outcome. Ann. Surg. 259 (2), 381–387.

## Chapter 18

1. The only ALS drug approved by the US Food and Drug Administration at the time of this writing is riluzole, which unfortunately extends survival by only several months. The underlying mechanism of likely drug efficacy was first identified in rodent models of motor neuron damage (Dib, M., 2003. Amyotrophic lateral sclerosis. Progress and prospects for treatment. Drugs 63 (3), 289–310).

2. The original Veterinarian's Oath adopted by the American Veterinary Medical Association (AVMA) House of Delegates in 1956 read: "Being admitted to the profession of veterinary medicine, I solemnly dedicate myself and the knowledge I possess to the benefit of society, to the conservation of our livestock resources and to the relief of suffering of animals. I will practice my profession conscientiously with dignity. The health of my patients, the best interest of their owners, and the welfare of my fellow man, will be my primary considerations. I will, at all times, be humane and temper pain with anesthesia where indicated. I will not use my knowledge contrary to the laws of humanity, nor in contravention to the ethical code of my profession. I will uphold and strive to advance the honor and noble traditions of the veterinary profession. These pledges I make freely in the eyes of God and upon my honor." http://clevengerscorner.com/column_details.php?topic=20110206. The Oath was revised in 1969 (and used at my graduation), then revised again in 2010 by the AVMA Executive Board to highlight animal welfare as a priority by the veterinary profession, reading as follows: "Being admitted to the profession of veterinary medicine, I solemnly swear to use my scientific knowledge and skills for the benefit of society through the protection of animal health and welfare, the prevention and relief of animal suffering, the conservation of animal resources, the promotion of public health, and the advancement of medical knowledge. I will practice my profession conscientiously, with dignity, and in keeping with the principles of veterinary medical ethics. I accept as a lifelong obligation the continual improvement of my professional knowledge and competence." https://www.avma.org/News/JAVMANews/Pages/x110101a.aspx.
3. Fox, M.W., 2006. Principles of veterinary bioethics. JAVMA 229 (5), 666–667.
4. The phenomenon of "compassion fatigue" has been well documented amongst human health care workers, and more recently encompassed those working in animal shelters and veterinary practices. One survey of veterinarians not engaged in laboratory animal medicine showed that almost one-third of the practitioners interviewed were at significant risk of acquiring compassion fatigue due to problematic clients and the business aspects of the practice (Figley, C.R., Roop, R.G., 2006. Compassion fatigue in the animal care community. The Humane Society Press, Washington, DC).
5. This is not to denigrate any religion, ethnic group, or nationality but there certainly are differences in how animals are valued in various countries, and societies with consequent variations in sensitivities about the care and use of laboratory animals. Having said that, each animal research facility I have visited in China invariably had a memorial garden on its grounds dedicated to the animals used in experimentation there. By contrast, obvious and poignant memorials such as these are rare in the United States.
6. Fox, J.G., Bennett, T.B., 2015. Laboratory animal medicine: historical perspectives. In Laboratory Animal Medicine, third ed. Fox, J.G., Anderson, L.C., Otto, G.M., Pritchett-Corning, K.R., Whary, M.T. (Eds.), Elsevier, London.
7. The common necessity of having to euthanize a high proportion of genetically engineered mice that can not be used and before they are weaned is well described by Hal Herzog in Chapter 8 of his book, "Some We Love, Some We Hate, Some We Eat: Why It's So Hard to Think Straight About Animals" (Harper, 2010. New York).
8. Birke, L., Arluke, A., Michael, M., 2007. The Sacrifice: How Scientific Experiments Transform Animals and People. Purdue University Press, West Lafayette, IN.
9. Scotney, R.L., McLaughlin, D., Keates, H.L., 2015. A systematic review of the effects of euthanasia and occupational stress in personnel working with animals in animal shelters, veterinary clinics, and biomedical research facilities. JAVMA 247 (10), 1121–1130.

# Chapter 19

1. Everyone understandably restricts the tabulation of laboratory animals to vertebrates because they are the animals with which the public is most familiar, in addition to being the most commonly regulated (cephalopods being the one group of invertebrates included by many institutions and oversight bodies due to their higher intelligence). However, I've argued for years that if we truly want an accurate count of all animals used in research, testing, and education, we need to include all invertebrates. If we do, then the number of nematodes (*Caenorhabditis elegans*) and fruit flies (*Drosophila melanogaster*) likely dwarfs all the others combined, because of their popularity for studying many fundamental biological phenomena such as aging, metabolism, behavior, and genetics. I have no idea how one would go about generating a representative tally of these two species, but it would be illuminating when playing the numbers game. On the other hand, given their biologically primitive state plus the high level of science illiteracy among the public, would anybody care?

2. Rowan, A.N., Loew, F.L., 2000. Animal research: a review of developments, 1950–2000. In: Rowan, A.N., Salem, D.J. (Eds.), The State of the Animals: 2001. Humane Society Press, 2001; USDA Annual Report Animal Usage: Fiscal Year 2016. https://www.aphis.usda.gov/animal_welfare/downloads/reports/Annual-Report-Animal-Usage-by-FY2016.pdf.

3. The Home Office of Her Majesty's Government of the United Kingdom, 2016. Annual Statistics of Scientific Procedures on Living Animals. https://www.gov.uk/government/uploads/system/uploads/attachment_data/file/537708/scientific-procedures-living-animals-2015.pdf.

4. Speaking of Research. Animal research statistics. https://speakingofresearch.com/facts/animal-research-statistics/.

5. Rowan, A., 2014. "Animal Testing, Animal Research & Alternatives: A Future Vision." Animal Matters Series Lectures, Tufts Cummings School of Veterinary Medicine, Grafton, MA, March 14 (presentation slides kindly provided by the speaker). Some readers may wonder about annual reports published by the Canadian Council on Animal Care (CCAC) that encompass all vertebrates used in research, testing, and education by institutions holding CCAC certification, as an additional comparator for animal use trends. In contrast to trends in the United States for species covered by the AWA and statistics from Great Britain, the CCAC annual reports indicate little change in total numbers from the mid-1980s to 2012 for most of those species that also fall under the AWA in the United States (http://ccac.ca/en_/facts-and-figures/animal-data/annual-animal-data-reports; special thanks to Andrew Rowan for sharing his tally of CCAC animal reports from 1975 to 2012). However, data in CCAC annual reports are not adjusted for variation in the number of reporting institutions. So it's unclear if a consistent number of animals reported over the years is a result of truly consistent use across Canada or a decline in total animal numbers masked by an expanding number of CCAC-certified institutions that weren't submitting data in earlier years. Also worth noting is that CCAC-certified entities include many agricultural and veterinary research and education institutions that for the most part would not be USDA-registered if they were in the United States.

6. Committee on Gene Drive Research in Non-Human Organisms: Recommendations for Responsible Conduct; Board on Life Sciences; Division on Earth and Life Studies; National Academies of Sciences, Engineering, and Medicine, 2016. Gene Drives on the Horizon: Advancing Science, Navigating Uncertainty, and Aligning Research with Public Values. National Academies Press, Washington. Available at: http://www.nap.edu/23405.

7. World Health Organization, 2016. 10 facts on malaria. http://www.who.int/features/factfiles/malaria/en/.

8. Shaw, J., 2016. Editing an end to malaria? Harvard Magazine. http://harvardmagazine.com/2016/05/editing-an-end-to-malaria.

9. Boodman, E., December 7, 2015. With potential to save human lives, CRISPR already sparing mice. STAT. https://www.statnews.com/2015/12/07/crispr-mouse-models/.
10. Reardon, S., 2016. Welcome to the CRISPR zoo. Nature 531 (7593), 161–163.
11. US Department of Health and Human Services. Organ Procurement and Transplantation Network. https://optn.transplant.hrsa.gov/.
12. Belluck, P., January 25, 2016. Monkeys built to mimic autism-like behaviors may help humans. New York Times. https://www.nytimes.com/2016/01/26/health/autism-genetically-engineered-monkeys.html?mcubz=1&_r=0.
13. Clayton, J.A., Collins, F.S., 2014. NIH to balance sex in cell and animal studies. Nature 509 (7500), 282–283.
14. For Zika virus, see Rossi, S.L., Vasilakis, N., 2016. Modeling Zika virus infection in mice. Cell Stem Cell 19, 4–6; Dudley, D., et al., 2016. A rhesus macaque model of Asian-lineage Zika virus infection. Nature Comm. 7, 12204. https://doi.org/10.1038/ncomms122. For Ebola virus, an interesting change of perspective in Ebola animal models arose after the 2014–16 outbreak in West Africa. For decades prior, when Ebola infection was believed to be mostly fatal, it was thought necessary to infect NHPs with a reliably lethal dose in order to mimic the human disease. That meant lots of monkeys were destined to die from hemorrhagic fever, including most or all that were in untreated/negative control groups in a given experiment (for an estimate of the number of monkeys involved, see Aldous, P., July 7, 2015. The silent monkey victims of the war on terror. BuzzFeed News. https://www.buzzfeed.com/peteraldhous/the-monkey-victims-of-the-war-on-terror?utm_term=.paoW6AvLR9#.lbAWa-NYg84). But after the 2014–16 outbreak in which the human case fatality rate averaged 40% rather than prior outbreaks with 80–90% deaths, Ebola researchers were forced to reconsider their earlier assumptions. They had to figure out how to administer non-lethal doses as well as extended supportive care to NHPs in order to model Ebola infection outcomes more accurately. This is mentioned only to remind oneself that any good science, including but not limited to animal research, benefits from frequent, evidence-based challenges to its convictions and practices.
15. Belluck, P., April 3, 2013. Dementia Care Cost Is Projected to Double by 2040. New York Times. http://www.nytimes.com/2013/04/04/health/dementia-care-costs-are-soaring-study-finds.html?mcubz=1.
16. US National Institute of Environmental Health Sciences, 2004. A National Toxicology Program for the 21st Century. Available at: https://ntp.niehs.nih.gov/ntp/about_ntp/ntpvision/ntproadmap_508.pdf.
17. National Toxicology Program, US Department of Health and Human Services. Tox21. https://ntp.niehs.nih.gov/results/tox21/index.html.
18. Anon., June 13, 2015. Towards a body-on-a-chip. Economist; Huh, D., 2015. Engineering human organs onto a microchip. TEDxPenn. https://www.youtube.com/watch?v=sCEWiFwWbXg&feature=youtu.be; Xiao, S., et al., 2017. A microfluidic culture model of the human reproductive tract and 28-day menstrual cycle. Nature Commun. 8, 14584. https://doi.org/10.1038/ncomms14584.
19. Willyard, C., 2015. Rise of the organoids. Nature 523, 520–522.
20. Capulli, A.K., Tian, K., Mehandru, N., Bukhta, A., Choudhury, S.F., Suchyta, M., Parker, K.K., 2014. Approaching the in vitro clinical trial: engineering organs on chips. Lab. Chip 14 (17), 3181–3186; Benam, K.H., et al., 2015. Engineered in vitro disease models. Ann. Rev. Pathol. Mech. Dis. 10, 195–262; Forum on Neuroscience and Nervous System Disorders; Board on Health Sciences Policy; Health and Medicine Division; National Academies of Sciences, Engineering, and Medicine, 2017. Therapeutic Development in the Absence of Predictive

Animal Models of Nervous System Disorders: Proceedings of a Workshop. National Academies Press, Washington. Available at: https://www.nap.edu/catalog/24672/therapeutic-development-in-the-absence-of-predictive-animal-models-of-nervous-system-disorders.

21. Grimm, D., 2016. From bark to bedside. Science 353 (6300), 638–640.

22. Davies, G.F., et al., 2016. Developing a Collaborative Agenda for Humanities and Social Scientific Research on Laboratory Animal Science and Welfare. PLoS one. https://doi.org/10.1371/journal.pone.0158791.

23. Panksepp, J., Panksepp, J.B., 2013. Toward a cross-species understanding of empathy. Trends Neurosci. 36 (8), 489–496.

24. Tennyson, A., 1850. In Memoriam A. H. H. Available at: http://www.online-literature.com/tennyson/718/.

25. Morgan, D.W., 2007. Whips of the West. An Illustrated History of American Whipmaking. Cornell Maritime Press, Centreville, MD.

26. Jepsen, T.C., 2000. My Sisters Telegraphic: Women in the Telegraph Office, 1846–1950. Ohio University Press, Athens, OH; Nonnenmacher, T. History of the U.S. telegraph industry. Economic History Association. https://eh.net/encyclopedia/history-of-the-u-s-telegraph-industry/.

27. Hu, C.K., Hoekstra, H.E., 2017. *Peromyscus* burrowing: A model system for behavioral evolution. Semin. Cell Dev. Biol. 61, 107–114; Bendesky, A., Kwon, Y.M., Lassance, J.M., Lewarch, C.L., Yao, S., Peterson, B.K., He, M.X., Dulac, C., Hoekstra, H.E., 2017. The genetic basis of parental care evolution in monogamous mice. Nature 544 (7651), 434–439.

28. Committee on Future Biotechnology Products and Opportunities to Enhance Capabilities of the Biotechnology Regulatory System; Board on Life Sciences; Board on Agriculture and Natural Resources; Board on Chemical Sciences and Technology; Division on Earth and Life Studies; National Academies of Sciences, Engineering, and Medicine, 2017. Preparing for future products of biotechnology. National Academies Press, Washington. Available at: https://www.nap.edu/catalog/24605/preparing-for-future-products-of-biotechnology.

29. Committee on Forecasting Future Disruptive Technologies; Division on Engineering and Physical Sciences; National Research Council; National Academies of Sciences, Engineering, and Medicine, 2009. Persistent forecasting of disruptive technologies. National Academies Press, Washington. Available at: http://www.nap.edu/catalog/12557.html.

30. To this last point, I had the pleasure years ago of attending AquaVet II, a 2-week course on the pathology of marine animals of commercial or research interest held at the Marine Biological Laboratory in Woods Hole, Massachusetts. While there, I met an accomplished turtle veterinarian at an inevitable evening gathering at the local tavern. I asked him what he would love to know about sea turtles that's a mystery but could help save those animals from extinction. He smiled and said sea turtles have an extraordinary ability to withstand infections that arise from boating accidents or unsuccessful attacks by sharks that would overwhelm other vertebrates; that they seemingly "refuse to die." I immediately proposed a research project that would obtain tissue samples from injured turtles recuperating at established rescue sites. These samples could be used to characterize the extent of the damage and determine which microbes may be present and contributing to infection. Subsequent blood or wound samples from the same patients could be analyzed to determine which haplotypes of genes are active at different stages of healing and how their gene expression patterns change over time. All of these findings may provide new insights on innate immunity and wound healing that would help not only injured sea turtles but also the rest of us. No further harm would have come to the injured turtles, no naïve turtles would be experimentally (intentionally) injured, and the research would engage experts from multiple professional disciplines who don't usually cross-talk. That research project is still waiting for someone to organize it.

# Index

Printed in the United States
By Bookmasters